c|net Do-It-Yourself HOME NETWORKING PROJECTS

About the Author

Jim Aspinwall is the author of six books covering topics from introduction to PCs through troubleshooting, configuration references, wireless networking, and PC hacking and mods. From 1995 through 2001 he satisfied the how-to appetite for PC and Windows users as the Windows Advisor for *Computer Currents*, and the Windows Helpdesk for CNET. In addition to writing about technology (as a hobby, BTW), Jim is an Extra Class amateur radio operator enjoying mountain-top radio system building and tower climbing. He volunteers his time as a Community Emergency Response Team (CERT) instructor and team lead within the City of Campbell, California, and as an Auxiliary Communications Service officer in the Coastal Region for the California Governor's Office of Emergency Services. During business hours (and after) Jim is responsible for the evaluation, configuration, and deployment of PC technologies for over 4800 users worldwide. He's a regular go-to guy when it comes to being prepared and jumping to action for technology or natural disasters.

c|net Do-It-Yourself
HOME NETWORKING
PROJECTS

24 cool things you didn't know you could do!

Jim Aspinwall

Mc
Graw
Hill

New York   Chicago   San Francisco
Lisbon   London   Madrid   Mexico City
Milan   New Delhi   San Juan
Seoul   Singapore   Sydney   Toronto

The McGraw·Hill Companies

Library of Congress Cataloging-in-Publication Data

Aspinwall, Jim.
  CNET Do-it-yourself home networking projects :
24 cool things you didn't know you could do! / Jim Aspinwall.
    p. cm.
  ISBN 978-0-07-148662-0 (alk. paper)
  1. Home computer networks. I. CNET (Firm) II. Title. III. Title:
Do-it-yourself home networking projects.
TK5105.75.A87 2008
004.6'8—dc22

                                              2007048291

McGraw-Hill books are available at special quantity discounts to use as premiums and sales promotions, or for use in corporate training programs. To contact a representative, please visit the Contact Us pages at www.mhprofessional.com.

**CNET Do-It-Yourself Home Networking Projects:
24 cool things you didn't know you could do!**

1234567890   QPD QPD   01987

ISBN   978-0-07-148662-0
MHID   0-07-148662-3

Sponsoring Editor Judy Bass	**Copy Editor** Bill McManus	**Illustration** International Typesetting and Composition
Editorial Supervisor Janet Walden	**Proofreader** Francesca Ferrie	**Art Director, Cover** Jeff Weeks
Project Manager Harleen Chopra, International Typesetting and Composition	**Indexer** Claire Splan **Production Supervisor** Jean Bodeaux	**Cover Designer** Jeff Weeks
Acquisitions Coordinator Rebecca Behrens	**Composition** International Typesetting and Composition	

To Kathy, with love

Contents

Part I   Easy

Part II   Advanced

Foreword

A few years ago you were considered a SUPER geek if you had a home network. It meant you were probably a computer hobbyist with Ethernet cables strung down the hallway in pursuit of some arcane form of machine-to-machine connectivity.

Not anymore.

Today's home network revolution was built on the "Three E's"—ease, entertainment, and elegance.

Ease of installation meant networks became at least comprehensible if not quite "plug and play," and the ease with which they share a broadband connection in the home is like magic. Entertainment is supercharged by a network's ability to bring all the audio and video in a home together for enjoyment in any room. And the elegance of Wi-Fi utterly changed the game; running wires through walls and under carpets had kept most of us at bay.

In this book we start with some fundamentals like connection sharing and printer sharing. These are the bread-and-butter parts of a network and gratifying to use every day. A little more advanced are projects that involve powerful security settings that lie within the network router, and setting up servers that live on the home network. You'll also find a constant TV-watching thread throughout, as we show you how to master and extend products like TiVo and Slingbox on your home network.

Home networks don't seem like such a technological luxury any more. They do such useful and indispensable services it's easy to imagine a day when they are a standard technology in almost every home. Until then, this book will keep you ahead of the curve.

Brian Cooley
CNET Editor-at-Large

Acknowledgments

Of course, special thanks go to Judy Bass at McGraw-Hill for yet another wonderful project, and for all of her support. I'm quite flattered to again work with my favorite editor, and coincidentally enough, on a collaborative project for CNET for whom I was the Windows Helpdesk columnist a while back—the "tech-family" bond is comforting. Steve G. at Verizon Wireless, and Brian J. at SlingMedia, both deserve special recognition and thanks for their help with evaluation product and support. Not in the least, to my co-workers and the hundreds, perhaps thousands, or even millions of users and readers I hope I've helped with each e-mail, article, and book—without them there would be no reason to share all this technology and empowerment.

Introduction

You may be thinking that home networking is so ubiquitous there could not possibly be anything new or different to do with this interconnected marvel of now everyday technology. You can surf the Web on any of two to three PCs, upload and download photos, fill an MP3 player to the brim, and chat with your friends and family via IM—what more could you want?

I've learned from a lot of my friends and co-workers that indeed, that's about all they can do with their home network now. But they also know about file sharing, printer sharing, wireless, voice chats, and webcams, and think they'll set up those "pretty soon"—until they hear someone else has hit a snag in their efforts to do more, and then they stop.

This book and the projects in it are about GO! You'll find out that you'll want to correct things in your firewall/router, Windows configuration, secure your network and computers, share that printer, get that wireless working in the back-forty (OK, back-forty square-feet, it's a small place, work with me, please...). You'll be sharing music, checking your own weather, watching TV on your laptop (*without* a tuner card), and enjoying numerous other aspects of home networking—reliably, and a lot more secure than before.

This is the book you will take to your friends and say, "hey, I found out how to..." and then be able to sit down at a computer and *do* "it" in 20–30 minutes. (Sorry, this book will not help you if you've forgotten your password and locked yourself out of your computer.)

What You'll Do

This book consists of 24 separate, but increasingly interrelated, projects covering the basics of broadband Internet connections through expanding your network in multiple ways, and taking advantage of a reliable and secure network to have some fun. Here are some examples of what you'll find:

- Six projects cover creating, securing, and expanding your home network from broadband through wireless.

- Six projects cover enhancing and sharing services between your Windows computers—from files to printers to remote control.

- Eight projects cover multimedia services, such as sharing weather and streaming media, to accessing resources inside your home network across the Internet.

- Four projects cover alternatives to conventional broadband, and configuring your network for general Internet services.

Some of the projects build on and refer you to related projects so you get the basics down and can move on to more advanced networking features. Each project starts with outlining what you will need, if any costs are involved, and a description of the project. For most projects you already have what you'll need—a broadband Internet connection, a router/firewall, and a computer or two or three.

 Visit this book's web page at http://diynetwork.cnet.com to see related videos.

Part I

Easy

Basic DSL Broadband Connection

What You'll Need

- DSL modem: Typically supplied by your service provider
- DSL filters: One for each voice telephone, fax, and modem
- Ethernet network cable: Usually included with your modem
- Cost: DSL Internet service $15–60/month; Filters $2–5; Cable $6–20; Network card $20

Beginning to explore the Internet with high-speed/broadband access is exciting, and is much more satisfying if done correctly from the beginning. This project will get you started making the new connections you need to enjoy the Internet with some serious speed.

DSL modems come in several varieties—from a basic just-a-DSL modem, to units with built-in firewalls, routers, Ethernet switches, and Wi-Fi access points. If you get a basic modem package, the setup is simple, can be expanded on, and typically requires some special setup at your PC—which is what we'll cover in this project. We'll add to your basic service in subsequent projects.

Step 1: Adjust Your Phone Connections

Your new DSL Internet connection requires some minor recabling of your existing phone connections. This doesn't mean you need to install new wires all over your house or apartment—you just have to add a DSL filter (see Figure 1-1) at each phone connection. When you subscribe to DSL service, the digital signal of DSL is mixed in with your normal voice service somewhere down the lines a few blocks away. This signal produces a rushing sound—like a fast-moving waterfall—if you listen to your phone without the DSL filters. Adding a DSL filter to each phone connection ensures that the annoying rushing sound is blocked from your voice phones, fax machine, and dial-up modems.

Figure 1-1

DSL filters block the broadband digital signals from interfering with your voice calls.

Your DSL modem must "hear" the rushing sound, but it is not disturbed by your voice, fax, and modem. You must connect the DSL modem directly to the phone line without filters. This is often accomplished with a DSL filter and phone splitter, shown in Figure 1-2, which provides one connection for your phone and another specifically for your DSL modem.

Figure 1-2

A DSL splitter makes it easy to connect your DSL modem at an existing phone connection.

To start things off, go to each of your voice phones and install a DSL filter at each connection. Most DSL modem packages provide two filters, though you may need more of them—they are readily available at electronics and even home improvement stores.

Step 2: Connect Your DSL Modem

When you get to the location where you plan to connect your DSL modem, install the DSL splitter between the phone jack and DSL modem, using the DSL jack on the splitter. Connect your telephone or modem to the "Phone" jack on the splitter. The connections for a DSL modem are shown in Figure 1-3.

Figure 1-3

A basic DSL modem has three simple connections—DSL phone line, Ethernet connection, and power.

Performing Steps 1 and 2 results in a connection configuration similar to the schematic shown in Figure 1-4—a main phone connection distributed to multiple phone jacks, and a DSL filter or splitter between the jacks and your phones or modem.

Figure 1-4

A typical DSL connection schematic includes multiple phone jacks, a DSL splitter, and DSL filters.

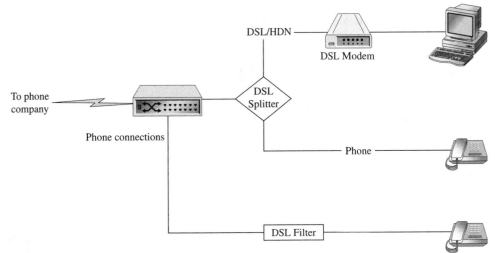

It may be possible to simplify your connections by using a single DSL splitter/filter—the filter side of the splitter for multiple phone jacks, and the DSL side of the splitter feeding a dedicated line for only the DSL modem, as shown in Figure 1-5. This technique requires separate phone line wiring to the location of your DSL modem, but requires only one phone connection change. You need to have access to your main phone connection point and be familiar with all of your phone wiring to make this technique work.

Figure 1-5

A simplified DSL connection schematic requires only one DSL splitter for multiple phone jacks, and a dedicated line to the DSL modem.

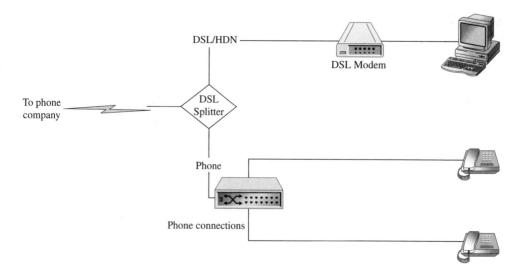

Step 3: Start Your DSL Modem

Your next step is to apply power to your DSL modem. Many DSL service providers recommend that you connect the DSL line and apply power to your modem up to ten days in advance of your DSL service activation date, to allow for line testing. If you are lucky, your DSL service may become active ahead of schedule.

Step 4: Connect the DSL Modem to Your Computer

You have one final connection to make—the network connection between the DSL modem and your PC, as shown in Figure 1-6. Assuming that your PC includes a built-in network adapter and Ethernet connection (most do, as do laptops), you only need to plug in an Ethernet cable between the modem and PC.

Figure 1-6

A typical DSL modem to PC network connection with Ethernet cable

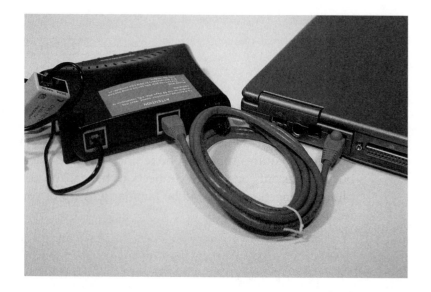

caution *Do not try to plug the Ethernet cable into a phone or modem jack, as it will not fit. A typical 2- or 4-wire phone connector will fit into an Ethernet jack, but doing so could damage the connector's pins or, worse, place excessive voltage onto the electronics of the Ethernet connection, which could cause the Ethernet components to fail.*

Your computer may detect and indicate this connection, if you have a network icon in your Windows tool tray, but you will not have a connection to the Internet without installing and configuring the software from your DSL provider.

note *Your DSL modem may provide only a 10 megabits per second (Mbps)/10BaseT or 100 Mbps/ 100BaseT Ethernet connection, which should work just fine even if your PC has a Gigabit Ethernet port. Not to worry—you will not be cheated out of high-speed performance. Your DSL connection speed will be only 6 Mbps or less, which "fits" just fine in a 10 Mbps, 100 Mbps, or Gigabit Ethernet port.*

Step 5: Install the DSL Software

The trickiest choice of the new DSL service process is whether to use the software included with your modem or use your PC's built-in support for DSL—if you are using Windows XP or Vista. The good news about using the software your DSL provider includes is that it is almost foolproof and working with the DSL provider's support people may be a little easier than working with only your PC's configuration.

The bad news about using the provider's software is that it installs a very customized web browser and myriad other pieces of software that make things easier on the provider but much more awkward for you and your PC's performance. It is almost impossible to remove this software once it is in place, and such attempts often result in having to reinstall your entire operating system and your preferred application software.

If you choose to install the provider's software, follow the installation steps carefully—you may be able to deselect one or more of the software options. If you have that option, you can still get the provider's requisite connection and support software, but leave unchanged your Internet browser, instant-messaging application, music download service, and so forth.

If the available software options include virus protection, antispyware, or firewall software, make sure you do not already have such products installed. Installing more than one brand of each type of protection software wastes disk space, creates conflicts between programs, and could render your system or Internet connection unusable. If you do not already have firewall software (keeping in mind that the Windows XP and Vista firewalls do provide adequate protection), choose at least that option, if not also the virus protection software if you have none.

Once your DSL provider's software is installed, it will walk you through a set of connection and configuration steps, much like we covered in Steps 2, 3, and 4, and then it will test the connection to your modem and try to establish a connection to the Internet. At this point, you may be asked for a username and a password, which will be used to protect your connection from unwanted use, similar to how they are used for a dial-up connection. If the connection fails you should contact your DSL provider to verify that your service is ready and the circuit is properly configured—common problem with DSL service.

note *Your operating system may recognize your DSL modem and be able to use it like a dial-up modem, precluding the need for the software shipped with your modem. Some DSL modems can be configured through a web page interface to connect to the Internet independent of PC or Mac software. In either case you still need to provide a username and password to connect to DSL.*

Precautions and Options

Once your connection is up and running, you're on the Internet and all set to go. Well, one computer is set to go and, depending on your choice of protection software, you may be able to enjoy the whole Internet. You may eventually consider further options, such as sharing your connection with more than one computer, configuring a wireless connection, and adding better protection, all of which are covered in later projects.

As you can see, the basic modem provided with most DSL service options provides only one Ethernet connection. The modem and service itself provides only one IP address; this and the single Ethernet connection provide service to only one client device—typically your computer. The basic modem also provides no built-in firewall protection or routing options, which are useful or necessary for a variety of virtual private network (VPN, for a remote office), Voice-over-IP (VoIP), and other enhanced Internet use connections.

As for protection, when you enter the world of broadband Internet access, your PC truly becomes a small but quite vulnerable part of the entire Internet. Without virus protection and at least Windows XP's firewall, your PC is a sitting duck for viruses, spyware, and a variety of exploits that will leave your PC helpless and at the mercy of persons unknown from around the world. Besides the possibility of becoming a victim of identity theft or having your bank accounts hacked, the next worst fate is having your PC mysteriously become part of a spam or hacker network, which will lead to your service provider turning off your Internet service—you'll never know what hit you!

To get more out of your broadband connection, the enhancements in subsequent projects offer you options for connection sharing, wireless, and enhanced services.

Project 2
Go Wireless

What You'll Need

- A wireless access point or combination router
- A wireless network adapter for your computer
- Setup from Project 1
- Your computer and web browser
- Cost: Wi-Fi access point/router $40–80; Wi-Fi adapter $20–40

You were happy with one computer in the den with a DSL or cable Internet connection, until a new laptop computer showed up over the holidays so that you can keep in touch while traveling, and perhaps even surf the Web from the patio, family room, or kitchen. You're now a multicomputer household with a single computer connection, and you want to connect your new computer to the Internet from anywhere in the house.

You may be tempted to tackle Project 10 and start running wires everywhere, but if your house is a flattop with no attic area, built on a slab foundation with no crawl space or basement, or your landlord simply won't let you tear into the walls, then wireless is your only option. While "everyone else" has been enjoying wireless networking for years, it's new to you, so this project will help you join the wireless party.

note *Wireless networking is a great substitute for wires, or rather where you cannot run wires, but wireless cannot deliver the solid 100 Mbps or 1 Gbps network speed you may want for computer-to-computer file sharing or blazing fast gaming.*

Whether you are starting from Project 1 or Project 10, the essential components are pretty much the same. If you're adding a wireless router to Project 1, you should go through Projects 10 and 11 to understand and accomplish the steps involved in changing from a single-computer network to multiple computers with the router and firewall configuration. If you've been through Project 10 and 11, similar steps apply to adding a wireless router. This project covers setting up the wireless aspects of the router and your computers, replacing an existing router with one that adds wireless functionality—though you could as easily add a wireless access point to your existing router.

Figure 2-1 shows a selection of wireless components you could encounter: a combination router/firewall that provides four Ethernet connections plus wireless access, a wireless access point only (no router features), a PC Card wireless adapter for laptops, and a PCI card wireless adapter for a desktop PC. That's right—you can make your desktop wireless too! If your laptop does not have a built-in wireless adapter, you'll need a PC Card or USB wireless adapter.

Figure 2-1

Myriad Wi-Fi devices are available to expand your network and computer capabilities.

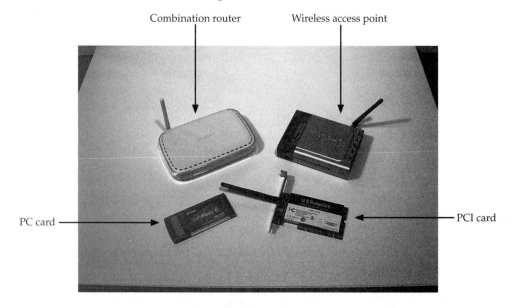

Adding wireless to your current network is a matter of replacing your current non-wireless router with a wireless unit, or adding a Wi-Fi access point to your present router, and spending about 15 minutes configuring the wireless unit to suit your preferences.

Step 1: Connect Your Wireless Router

The network schematic for this project is represented in Figure 2-2—the wireless router connects between the modem-to-computer connector, taking the place of the nonwireless router.

Figure 2-2

Wireless access allows many more flexible configurations for your local network.

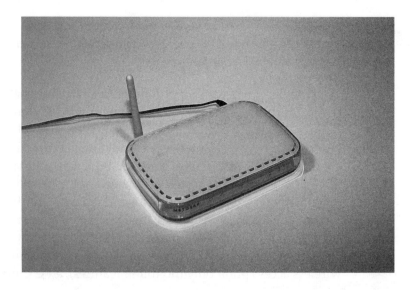

Follow this simple sequence to establish the connection between your computer and the router:

- Connect the Ethernet cable from your DSL or cable modem to the WAN/Internet port on the wireless router.

- Connect the Ethernet cable from your computer(s) to the LAN ports on the wireless router.

Step 2: Configure the Wireless Router

Before we get too concerned about going wireless, perform the router configuration as you would in Project 10, verify that your computer(s) can "dial up" and connect to DSL and establish Internet access, and then set up the wireless configuration. Following the instructions of the new router, you may have to change the IP address of one or more computers to access the configuration menus.

note *Although the passwords, menus, and some of the settings may appear different between the various brands of routers and firewalls, they all do the same things.*

Open your web browser, access the router's configuration menu, and locate the settings for wireless, as in the example shown in Figure 2-3. You are looking for the router's name, SSID name, and type of security to keep your network private.

Figure 2-3

The basic wireless configuration will get you going in the right direction.

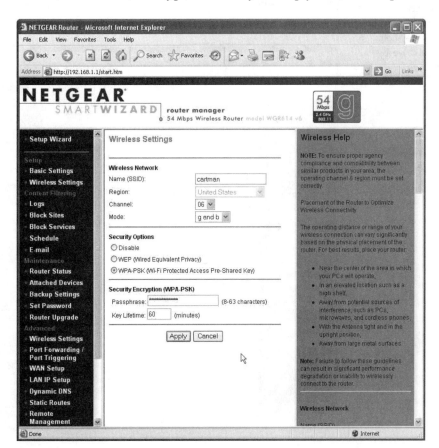

Many of us recognize the default names (SSIDs) of common Wi-Fi access points, and can expect that those named Linksys, NETGEAR, and so forth have not been secured by their owners, and thus are vulnerable to others taking advantage and gaining free use of the Internet access and bandwidth these access points provide. Change the name of your access point to something obscure, and perhaps ominous. Names such as "Infected" or "Reboot in progress" may wave off some folks, and certainly will not identify the manufacturer (and perhaps hidden vulnerabilities) of your equipment.

You also need to use the highest security measures (encryption of the wireless data streams) available. In the case of home wireless setups, Wi-Fi Protected Access (WPA) is significantly better than Wired Equivalent Privacy (WEP). Choose a difficult passphrase/passkey, preferably with more than six characters, including some numbers and symbols to mix it up. If you must use the name of your pet, like Scooter, substitute some numbers for the letters by making it into 5c00teR—and then remember how and what you obscured.

Simply changing the name (SSID) of the access point won't foil those looking for free bandwidth. Turning off the broadcast of your access point name (for example, clearing the two check boxes in Figure 2-4) is another way to at least temporarily

Figure 2-4

Wireless security settings cannot be overlooked—they help prevent casual drive-by bandwidth abuse.

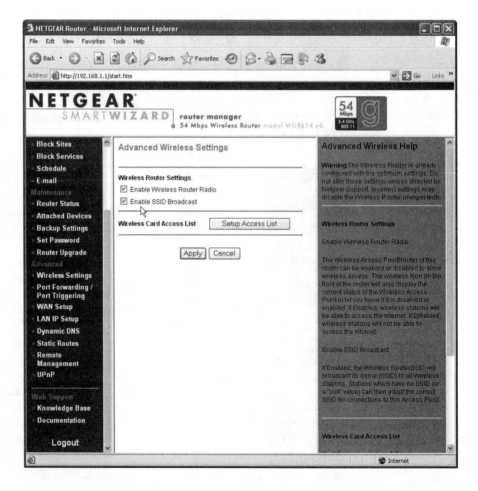

obscure your network from casual Wi-Fi opportunists (though you do need to remember the name when you set up your computers to access the network).

Save the settings you have changed and let your router restart (this may take up to five minutes). Then, you are ready to configure computers to use Wi-Fi for network and Internet access.

Step 3: Set Up Your Computer to Use Wireless

Whenever you encounter a new Wi-Fi location, you'll probably want to "hook up" and surf. Many public Wi-Fi spots provided by coffee shops and local communities require no configuration—merely double-click the hotspot's name and, cha-ching, you've got Internet gold. To get to that point, you can click Start, right-click My Network Places, click Properties, right-click your wireless network adapter's icon, and then select View Available Wireless Networks to get to the dialog box shown in Figure 2-5.

Figure 2-5

Windows' Wireless Network Connection dialog box showing nearby Wi-Fi access points

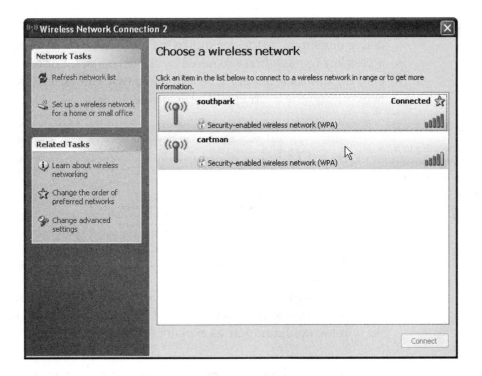

This dialog box is how Windows shows you the wireless scene around your computer. Dell, Intel, Cisco, and others have more robust and descriptive Wi-Fi connection programs, represented in Figure 2-6 for a Dell laptop.

Figure 2-6

One of Dell's Wi-Fi
configuration windows

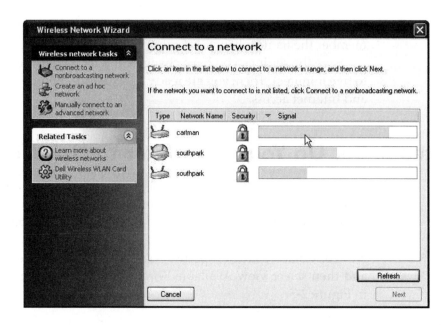

To "hook up" with one of the available networks, double-click its name to open a dialog box similar to the one shown in Figure 2-7, which prompts you for the network access key or passphrase required by the access point or router to let your computer onto the network. Simply type in the key twice and click OK, and in moments your network and the Internet will be available.

Figure 2-7

When connecting to a
secured Wi-Fi network
you are prompted for
the security key.

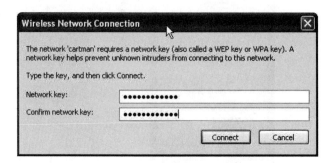

If your Wi-Fi network requires more intricate configuration, or you know of a Wi-Fi network that does not broadcast its name (SSID), you have to dig deeper into the wireless configuration, as explained in Step 4.

Step 4: Configure Your Computer for Hidden or More Secure Wi-Fi Systems

To add a new network that may not be broadcasting its SSID, or to set specific parameters for a known network in advance, you need to access the Properties dialog box for your wireless network adapter:

1. Choose Start | Connect To | Show All Connections to open the Network Connections window, shown in Figure 2-8.

Figure 2-8

Accessing the properties
for your wireless
network adapter takes
but a few clicks.

2. Right-click the name of the network connection, choose Properties to open
   the Wireless Network Connection Properties dialog box, and click the Wireless Networks tab, shown in Figure 2-9.

3. Select the wireless network adapter from the Preferred Networks list and
   click Add to open the Properties dialog box, shown in Figure 2-10.

Figure 2-9

The Wireless Networks
tab lets you manually
create new network
connections.

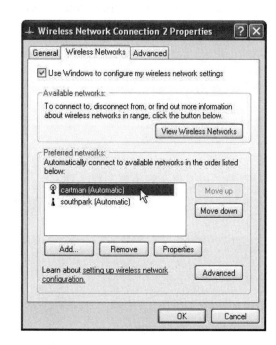

Figure 2-10

The Properties dialog box of a specific wireless network allows you to select the type of authentication and data encryption.

In the Properties dialog box, identify the network by name and complete the security property settings. As shown in Figure 2-10, the network (SSID) named cartman is protected by the WPA-PSK authentication method using TKIP to encrypt the data going back and forth. These are very cryptic designations for "more secure than not configuring any security settings."

Set the Network Authentication, Data Encryption, and Network Key values according to those required by the wireless network you will be using. Click the OK button to complete the configuration process. Once configured you can connect to the network through the available wireless connections dialog, (see Figure 2-5).

Project 3
File Sharing

What You'll Need

- **Your computers**
- **Your local network**
- **Cost: $0**

Local network file sharing displaced the "sneaker net" years ago, but if this is your first personal/home network, network file sharing may be met with the likes of homecoming-fanfare. To set up file sharing between computers—one computer holding the files you want others to have access to, without sending e-mail or copying to disks, CDs, or USB memory sticks—you need to decide and configure what to share, and give permissions to other users so they can access the shared drive or folder.

In the two previous projects, you set up your local network so that all of the computers are aware of each other as a unified workgroup, with you running around to every computer in the house configuring user accounts. With all of this in place, file sharing is a matter of deciding what to share, and with whom.

Step 1: Establish a Folder to Share

While it is possible, you should never share an entire disk drive, especially not the system drive, C:. The system drive contains the operating system and program files so often implicated in and infected by hacks, viruses, and malware.

Windows XP offers one method of sharing files and folders—Simple File Sharing. XP Home Edition offers only method of sharing, Simple File Sharing, as described here. Only XP Professional offers the ability to disable Simple File Sharing to allow more granular control over which users can see that you are sharing files, and whether or not they can read, write, modify, or delete files within the shared folder. The options and access to them are similar for Windows Vista.

Let's start with a computer named LT2. To provide for a single, private area to share files on LT2, create a new folder and share only that one folder. Open Windows Explorer (My Computer), select the C: drive, and then create a new folder—"Share" is a good enough name to identify the purpose of the folder, so use that for this project.

Open My Computer, then open the C: drive (or other drive you wish to share files from), and create a new folder to contain only those files you wish to share with others. There is virtually no limit to how many of these folders you can create, set up sharing for, and select the content and the users who have access to it. For instance, you may create folders for family photos, home budget, social events, school issues, and so on, and share them each differently with different users.

Step 2: Share Your "Share"

Right-click your new Share folder and select Sharing and Security, as shown in Figure 3-1. A Properties dialog box opens with the Sharing tab displayed, as shown in Figure 3-2. Check the Share This Folder on the Network check box and then provide a Share Name, which is the name you would like this folder to be known by across your network (I chose "lt-share" to identify the shared folder on a laptop). Click the Apply button.

Figure 3-1

Right-click the folder you will share with others and choose Sharing and Security.

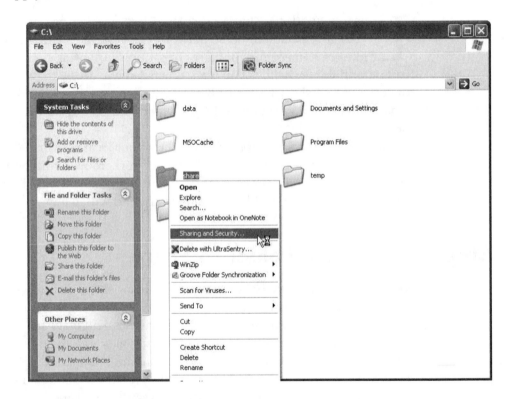

To allow users to change (modify or delete) or write new files in your share, within their respective user account rights, check the Allow Network Users to Change My Files check box. If this check box is cleared, other users may only read and copy the files from the share. Once this step is complete, your shared folder and files become "visible" on your network and are ready to be used by other users from other computers.

Figure 3-2

Providing a unique name to a selected folder to share, to identify the shared resource, and allowing others to modify the files within the share

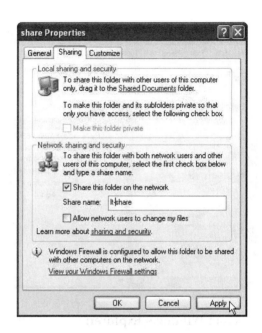

note   *The sharing settings are valid and active only when the user who established the sharing settings is logged on to the computer. If another user is logged on, the same folder is not shared, unless the logged-on user configures the shared resource as described in this step. This means that if user Jackie established a shared folder and files, other users can share those files while Jackie is logged on. If later user Tommy logs on to the same computer, the folder and files Jackie shared will not be shared with others, unless Tommy also configures that folder to be shared.*

Step 3: Connect to the Share from Other Computers

So far we have one computer named LT2 configured to allow others to see and access the C:\SHARE folder, over the network by other computers, when user Jim is logged on to the computer.

If we move over to work on another computer, say WILDFLOWER, and want to access the shared files on the computer named LT2 with the shared folder we called SHARE, we have two ways to navigate to the share—through the My Network Places window, or in Windows Explorer.

From My Network Places:

1. On your Desktop, double-click the My Network Places icon, or choose Start | My Network Places.

2. In the My Network Places window, shown in Figure 3-3, you should see all the computers and shares available on your network, including the shared folder on computer LT2.

3. Double-click the lt-share icon in the window to see all the files that are shared, as shown in Figure 3-4.

Figure 3-3

My Network Places is the place to see the shared resources available on your network.

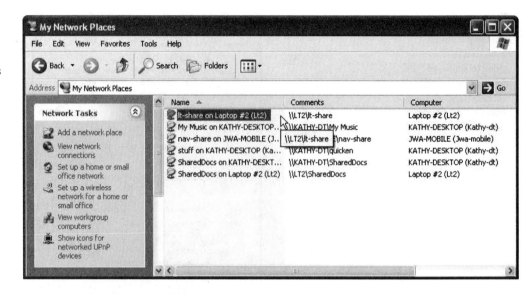

From Windows Explorer:

1. In the Windows Explorer Address bar, type in two back strokes followed by the name of the computer that has the shared resource you are seeking—in this case **\\lt2**, as shown in Figure 3-4—and then press the ENTER key or click the Go button.

2. Double-click the icon for lt-share to view the files we know we've shared and have access to.

Figure 3-4

Computer LT2 has several potential shared resources.

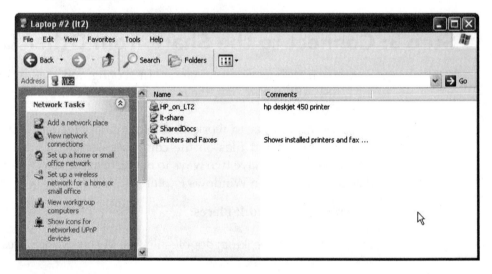

Navigating to a shared resource from My Network Places or Windows Explorer gets you to the desired resources. In Simple File Sharing mode, all users have equal access to the files present. You may wish to disable Simple File Sharing to have more granular (and complicated) control over file shares and what users can do with the shared resources. A general rule is, if you don't want others to read, copy, modify, write over, or delete your files, do not share them!

Project 4
Printer Sharing

What You'll Need

- Your computers
- Your printer(s)
- Your local network
- Cost: $0

As much fun as sharing files can be, I can think of nothing more useful to do with a local network than share a printer or two. After all, lugging a printer from one PC to another (along with paper, USB cable, power adapter and all) is pretty awkward. Sure, you can share files, put your files to be printed on the PC connected to your printer, and print that way, but why not skip a step or two and simply send your printer tasks to one or the other shared printer?

Step 1: Establish a Shared Printer

Much the same as sharing files, as covered in Project 3, sharing a printer requires that you establish the shared resource and make it available to other computers and users. At the computer attached to your printer:

1. Choose Start | Printers and Faxes.

2. In the Printers and Faxes window, right-click the icon of the printer you want to share and choose Sharing, as shown in Figure 4-1.

> **note** *It is important to share only printers that are supported by all of the operating systems on your computers. If you have a really old printer, or different operating systems on your computer, the wizard may not be able to find or install a driver for your printer onto a particular operating system. This may be your excuse to upgrade to a Vista-compatible printer!*

3. In the printer's Properties dialog box, on the Sharing tab, shown in Figure 4-2, select the Share This Printer radio button and then type in the Share Name field the name you would like this printer to be seen as on your network. In this example, we have an H-P DeskJet 450 to share from computer LT2.

Figure 4-1

Selecting a printer to
share and accessing its
share properties

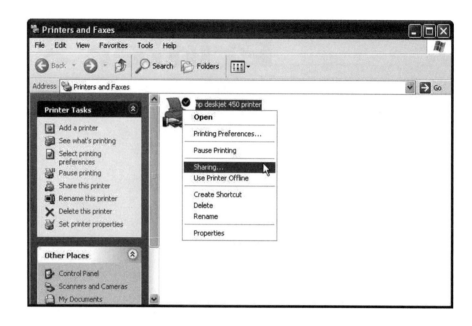

Figure 4-2

Configuring your
printer to be shared on
the network

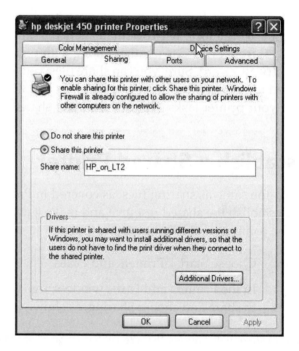

4. Click the OK button to close this dialog box, and your printer will be available on the network.

As with file sharing, this share is available only when the user that created it is logged on. To make sure this printer is shared when any user is logged on to this computer, you must log on as each of those users and re-create the share within each of their accounts.

Step 2: Connect To and Print on Your Shared Printer

To use a shared network printer, you add a printer just as you would a local printer directly attached to your computer. Instead of searching for a local printer, you browse the network:

1. Choose Start | Printers and Faxes.

2. In the Printers and Faxes window, click Add a Printer under Printer Tasks. This starts the Add Printer Wizard. Click the Next button.

3. The next dialog box, shown in Figure 4-3, gives you choices to look for a local or network printer. Choose the A Network Printer, or a Printer Attached to Another Computer radio button and then click the Next button.

Figure 4-3

To use the shared printer on another computer, you must select A Network Printer to find it.

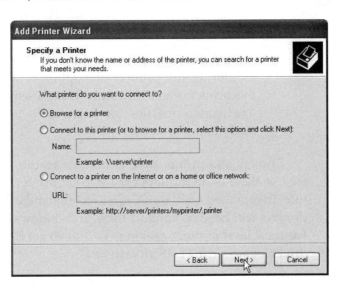

4. In the Specify a Printer dialog box of the Add Printer Wizard, shown in Figure 4-4, choose Browse for a Printer and then click the Next button.

Figure 4-4

Let Windows do the work of finding your printer by choosing to browse for a printer.

5. Windows presents a list of all shared printers found on your network, as shown in Figure 4-5. Select the printer you are looking for and then click the Next button.

Figure 4-5

Select the shared printer you want to use from the Shared Printers list.

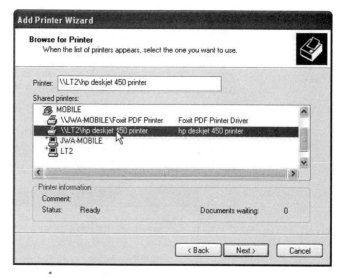

6. The Connect to Printer message box shown in Figure 4-6 appears, indicating that a driver for this printer must be installed. Click the Yes button to continue with the driver installation. If there is a driver available for this computer's operating system, it will be installed. The driver installation may take a minute or two. If no suitable driver can be found, you need to check the printer manufacturer's web site for an updated driver to support your operating system.

Figure 4-6

The Connect to Printer message alerts you to the need to have a printer driver installed.

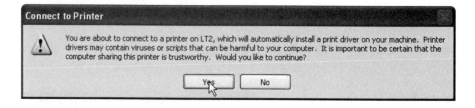

7. In the next dialog box, you are asked if you want to use the selected printer as your default printer—the option is up to you. The final result of adding this printer is an icon and label representing the printer atop the symbol for a network connection, as shown in Figure 4-7. At this point you're done and ready to print.

Printers and printer drivers are one feature that Windows XP supports pretty well, if there is driver support for your printer, or you can find a suitable substitute driver based on manufacturer recommendations. Many two- and three-year-old printers will have no support for or in Windows Vista, causing certain aggravation. For most laser printers, you may be able to get by with a generic HP PCL 5 or PCL 6 driver, or a generic PostScript driver.

Figure 4-7

A shared printer
appears on your
computers with
a network line
underneath the icon.

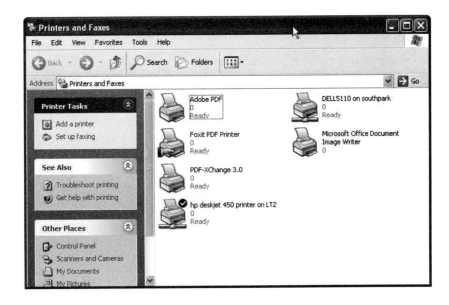

Shared Network Storage Without the Server

What You'll Need

- Ximeta or similar network storage adapter module and driver software
- Local area network
- Cost: Enclosure $40–60; Disk drive $100–300

For some reason, most of what we do with a computer isn't worth a whole lot unless we can share the results with family, friends, or coworkers. At home we usually want to share digital photos or music files. The evolution of file sharing has progressed from the original "sneaker net"—passing files back and forth on disk walked to and from people—to enhanced sneaker net with CDs and DVDs, to connecting computers with serial- or parallel-port data transfer cables, to sending files as e-mail attachments, to setting up local file sharing from one computer to the other (Project 3), and finally to building our own file servers.

Sending files as e-mail attachments seems kind of a silly thing to do, especially within the same household—it's horribly slow (worse if you use dial-up), and if your e-mail service limits the size of file attachments, you may not be able to share the file at all. Setting up sharing of local computer resources is fraught with security risks and requires both computers to be up—which is not very efficient or eco-friendly because of all the power consumed. Certainly you can build a file server—using your normal desktop operating system or a server OS—but this gets expensive because it consumes yet another PC and more energy, an additional software license, and the maintenance hassles of configuration and security. Unless you are well versed in networking, operating systems, and cross-platform data transfers, file sharing with local computers and servers typically only works between the same type of computer system: PC-to-PC, Mac-to-Mac, or Linux-to-Linux.

A very simple, inexpensive, and fast way to share files at home is with network-attached storage—disk drives that can be accessed directly over your local area network (LAN). Ximeta, Linksys, and other companies offer such devices (an example from

Ximeta is shown in Figure 5-1) as either complete network interfaces, including the disk drive, or as network-enabled drive enclosures that you install a disk drive into.

Figure 5-1

Ximeta's NetDisk network storage enclosure with a self-installed disk drive

Inside most USB, IEEE-1394/FireWire, and network-attached storage units is simply a normal IDE or SATA disk drive and a set of electronics to convert the external connection into the appropriate disk drive interface (see Figure 5-2).

Figure 5-2

The secrets of a network-attached storage drive are a small internal CPU board and a standard IDE or SATA disk interface, and a normal 3.5-inch disk drive.

Step 1: Connect the Storage Module

The easy part of installing your new shared network storage is connecting it to your network—simply attach the power cable and connect an Ethernet cable to your network switch, as shown in Figure 5-3. You will be able to format and configure the drive after you install the software for it.

Figure 5-3

Network drive module connected to network switch by Ethernet cable

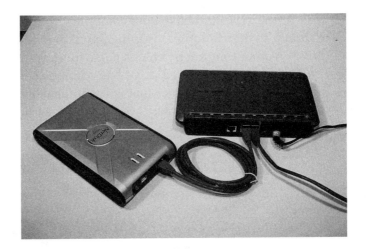

note Although these units usually have a USB connection so that you can connect them directly to a computer as a portable disk drive and format them for your operating system, it is best to start out configuring them as network storage so that your operating system's disk management drivers do not get confused when they encounter the new network drive. You can use the unit through the USB connection after configuring the drive on the network.

Step 2: Configure the Software

The Ximeta Netdisk network storage software and driver installation wizard takes you through four basic dialog boxes, and then a Restart dialog box to complete the process. A restart is necessary to apply the network device drivers and start the driver application you use to configure and make your drive accessible to the operating system.

After restart, a new icon appears in the notification toolbar at the lower right of your screen. Right-click the icon and select Register a New Device to open the NDAS Device Registration Wizard, shown in Figure 5-4. You are beginning the sequence of

Figure 5-4

Provide a name for your network drive in the NDAS Device Registration Wizard.

NDAS Device Registration Wizard

NDAS Device Name
Specify the name for the NDAS device.

Please choose a name for the NDAS device. This name will be used to identify the NDAS device in your system only.

The name can be up to 30 characters long.

Please enter the name.

Media

Click Next when you finish.

< Back   Next >   Cancel

making the drive known to your computer, from providing a name for the drive that is unique to this computer to saving the configuration to make it easier to register the drive on other computers. Enter a name for the device and click Next.

Each network-attached drive has a unique serial number to identify it on your network, and a key number that is used to enable write access (so you can actually put files on the drive). Enter these numbers, as shown in Figure 5-5, and then click Next.

Figure 5-5

Enter the unique numbers for your specific disk drive chassis to make it available to your computer.

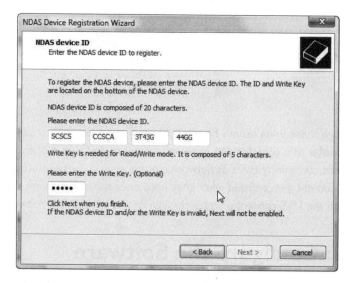

After the drive is registered, you have the option of mounting the drive to be written to and read from, to be read only (you cannot write files to the drive), or not yet mounted (does not appear as a drive in My Computer and is not available until mounted), as shown in Figure 5-6. When the drive is first installed, you must mount it with read/write access so that it can be formatted, so choose the "Yes, I want to mount this NDAS device as Read/Write Mode" radio button and click Next.

Figure 5-6

Select the Read/Write drive-mounting option so you can format your new disk.

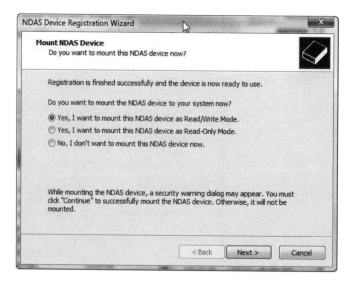

Figure 5-7

Completion of
the network drive
registration wizard with
the option to save the
drive configuration

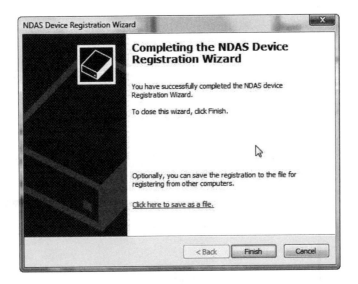

After you click the Next button, the software will attempt to mount the drive and
if successful displays the final step of the registration wizard, as shown in Figure 5-7.
Notice the text near the bottom of this dialog box and the Click Here to Save as a File
link. This is a feature that allows you to save this drive's registration information for
quicker registration on other computers. Click the link to open the Save As dialog
box, shown in Figure 5-8, and save this file to your local disk, a removable disk, or a
USB memory stick so that you can e-mail it or copy the file to your other computers—
PCs or Macs.

Figure 5-8

Saving the drive
configuration to a
file for use on other
computers

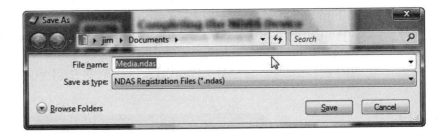

Step 3: Format Your New Drive

Formatting your new drive can be done with the Disk Management application under Windows XP or Vista. Click Start, right-click the My Computer icon, and select
Manage. In the Computer Management window, select Disk Management.

Scroll through the list of drives in the lower-right pane until you locate an entry
that shows an Unallocated disk. Right-click in the Unallocated disk area and select
New Partition, as shown in Figure 5-9, to start the New Partition Wizard. Click Next,
which takes you to the start of the parameters for partitioning and formatting the
drive.

Figure 5-9

Opening the New
Partition Wizard for
your new drive.

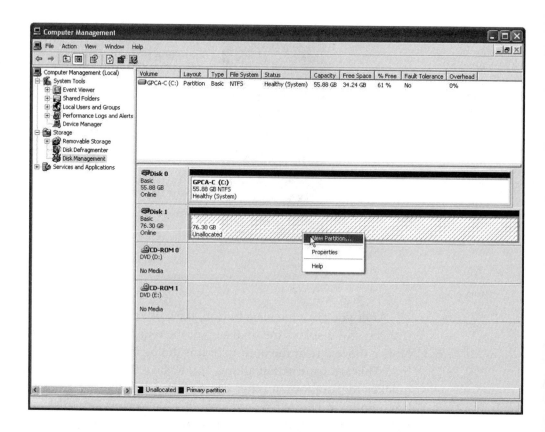

Leave the default Primary Partition selected and simply click Next. Accept the default partition size and then click Next to get to the drive letter selection, as shown in Figure 5-10. I suggest picking a drive letter beyond E: through K: or L: to allow "room" for multifunction card readers and USB sticks to obtain a nonconflicting drive assignment. Using a designation like M: for "multimedia" or S: for "shared" might make it easier to associate the drive letter with the purpose of the drive. Click Next.

Figure 5-10

Drive letter selection

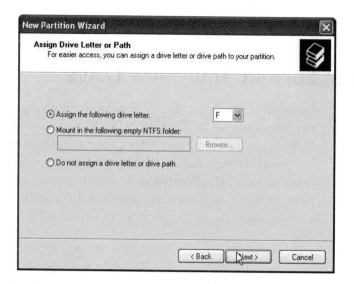

The next dialog box, shown in Figure 5-11, provides the final settings. Leave the File System type as NTFS (unless FAT32 is an available selection) and Allocation Unit Size parameter set to Default. Provide a name for the drive, click the Perform a Quick Format check box, and then click Next. In the Completing the New Partition Wizard dialog box, click Finish to begin the partition and format process. It can take from 5–15 minutes to format the drive, after which it will be ready for use.

Figure 5-11

Setting the final parameters to format your new drive

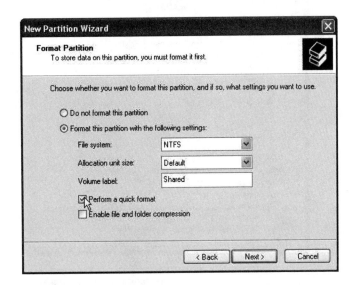

Step 4: Use Your Shared NetDisk on Other Computers

With your first network drive setup complete, you are ready to make the drive available to other computers. Follow Step 2 to install the network drive software and drivers and then restart the computer. After restart, you can register the drive manually, as illustrated in Step 2, or import the registration information from the configuration file you saved.

To use the configuration file you saved, copy it to the other computer(s) using a disk or USB stick, move the diskette or USB stick to another computer, open the drive in My Computer, then simply double-click the configuration filename (Figure 5-12). The net drive management software will "pick up" the file as in Figure 5-13, to allow you to register the NetDrive on this computer. Click the Register button and the shared drive will be "known" to this computer.

Once the drive is registered, all you need to do is mount the drive—in Read-Only or Read/Write mode—to be able to access it in Windows Explorer.

note *After your drive is installed and shared, if you manually register the drive for other computers, you may wish to eliminate the write key from the configuration so that others sharing the drive cannot add files to or delete files from the drive. To do this, right-click the NetDrive tool tray icon, then right-click the name of the NDAS device you want to remove the write-key from and select Properties. In the Properties dialog for the drive, next to the Write Key edit box click the Remove button. The computer you perform this action on will no longer be able to write to the network drive.*

Figure 5-12

Select the NDAS registration file for the NetDrive you want to use on this computer.

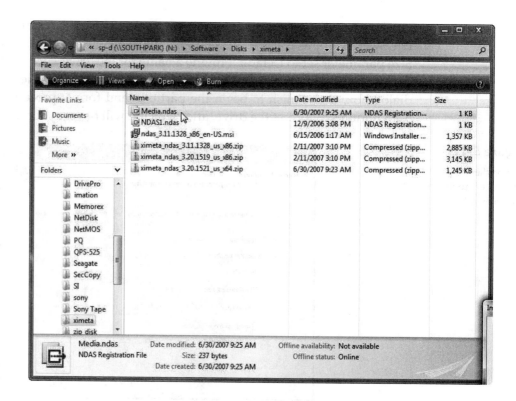

Figure 5-13

Register the NetDrive device on this computer to gain access to it.

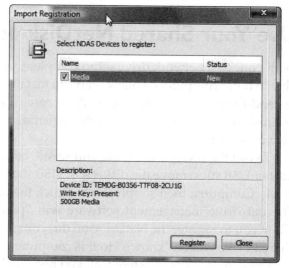

Step 5: Mount and Unmount Drives

To control whether a shared drive is accessible or writeable to your computer, right-click the icon in the system tray and select your desired mounting option, as shown in Figure 5-14.

note Note: It is a good practice to unmount the drive when it is not in use, especially if you are writing files to the shared drive. Leaving a drive mounted in Read/Write mode may prevent others from writing their files to the drive.

Figure 5-14

Network drive
management controls
are accessible through a
system tray icon.

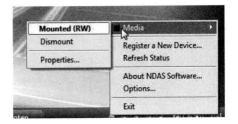

Webcam Server

What You'll Need

- A spare computer
- Your local network and Internet connection
- Webcam and driver software
- webcamXP webcam hosting software
- Cost: Webcam $20–80; Software $80

W hether for home security monitoring, keeping an eye on the nanny or cleaning crew, or watching the traffic pass outside your home, running a webcam and being able to view it while you are away can be pretty useful if not entertaining. You only need to add a camera and tweak your router's firewall settings to enjoy this project to its fullest.

Almost any webcam will work—as long as you can view yourself and capture snapshots on your local PC, you've met the minimum requirement to be able to use your webcam for security or entertainment. The toughest part may be finding a camera you like, or one that will support your operating system. I have three wonderful cameras, but had to buy a fourth to use with Windows Vista, 32- or 64-bit.

If your camera has a face-tracking feature, as do many of the Logitech cameras, you may want to turn off this feature until you get everything in the next few steps working correctly. Another caveat before moving on to the step that enables you to watch your camera on your local network or the Internet: after you have verified that your choice of camera works, turn off the camera software! Windows is not yet smart enough to allow you to run your camera's software and network/Internet camera software at the same time and have both provide the camera image you want all the time.

note *Canon, Panasonic, and Sony offer selections of steerable webcams with incredible picture quality and functionality. Since these products are in the $800-and-up price category, you will probably find that your $20–80 electronics store webcam from Creative or Logitech is more than adequate.*

Step 1: Install the webcamXP Software

I found webcamXP while looking for an easy-to-use program to handle multiple video sources—both PC cameras and video capture sources—as well as provide a built-in web server, e-mail picture or video captures on schedule, and upload pictures to web pages. Although I use the Pro version, you may find that the non-Pro version, at $30, works great for you. There is nothing tricky about installing webcamXP (see Figure 6-1), but the configuration has so many wonderful options that a little guidance will help you get things started.

Figure 6-1

The webcamXP software window provide access to all features.

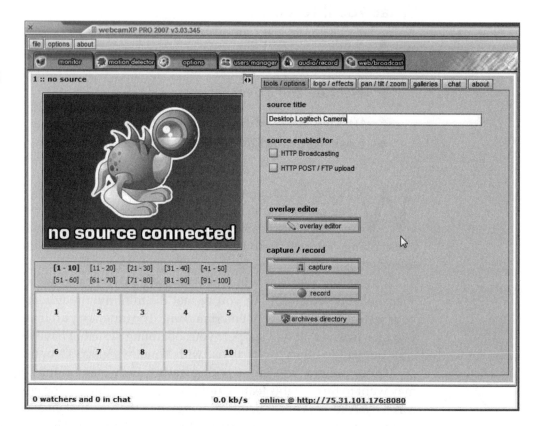

The first step is to select a video source, your webcam, so webcamXP has something to work with. Right-click the box labeled 1 and select Connect just under the camera icon to open a menu of video source types, as shown in Figure 6-2. You should be able to use the DirectX Video Source, which is generic to most Windows installations, then you will see your camera model listed—simply click the name of your camera as the desired video source.

A live image from your camera should appear in the box that was labeled 1 and in the main viewing area, as shown in Figure 6-3. The image in the main viewing area changes to the camera you select in the blocks below. webcamXP PRO can support up to 100 video sources.

Figure 6-2

webcamXP allows
you to choose from
a number of video
sources.

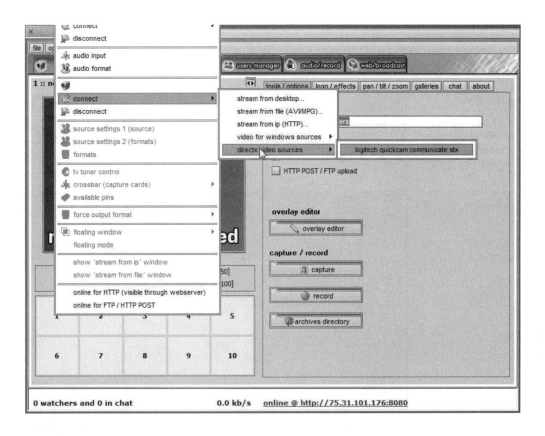

Figure 6-3

webcamXP displays
the live image from the
selected video source.

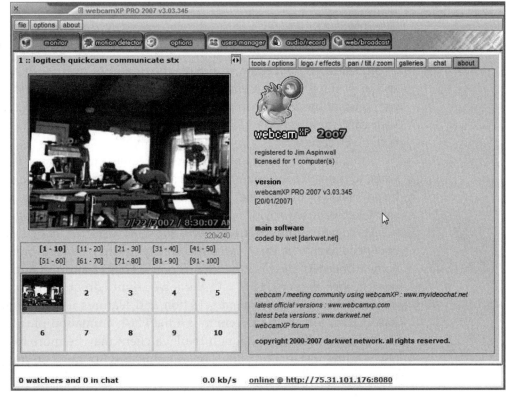

Click the Web/Broadcast tab, shown in Figure 6-4, to set up webcamXP to make your camera views available as web pages on your local network, and later over the Internet. By default, the program is already set to be its own web server.

Figure 6-4

Everything you need to broadcast and share your webcam is on one configuration page.

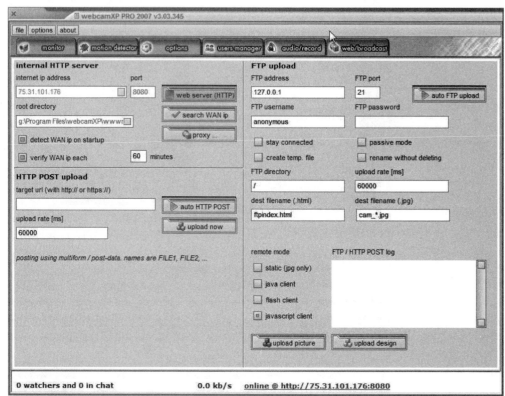

To share your webcam, you have the option of using webcamXP's internal web server, posting your camera output to a web page via HTTP POST (provided your web host allows you to do so and you provide the appropriate code), or uploading your camera output via FTP to an FTP download site or the file system on your external web server.

Using the web server option makes the camera images available on your local network, and if you configure your firewall in Step 3 to allow it, the image can be viewed by anyone over the Internet. I also upload one set of images every minute to my personal web page to provide a snapshot view and verify everything is working at home.

Make note of the port number webcamXP uses for its built-in web server—you may need to change this to accommodate your firewall configuration. To change this value, if needed, deselect the Web Server (HTTP) button, change the port, and then reselect Web Server (HTTP). In the lower-right corner of this page are settings for remote control of the camera selection and features using the built-in web server. Rather than force any viewers to use the Flash animation plug-in, I chose the ubiquitous JavaScript client, though the Java client may be more robust or faster, but it also requires installation of the Java Runtime Environment on computers viewing the webcam page.

Step 2: Configure Your Firewall for Webcam Access

In Project 11 we cover configuration steps for your router and firewall, and the latter steps apply to this project—we need to configure the firewall to allow access to our webcam server software from across the Internet.

Using as an example my ISP-provided router/firewall, shown in Figure 6-5, I can create specific "pinholes," otherwise known as redirection of inbound TCP/IP traffic to specific IP addresses/computers on specific TCP/IP ports.

Figure 6-5

Most router/firewall appliances allow you to be very specific about mapping external TCP/IP requests to individual computers and TCP/IP ports.

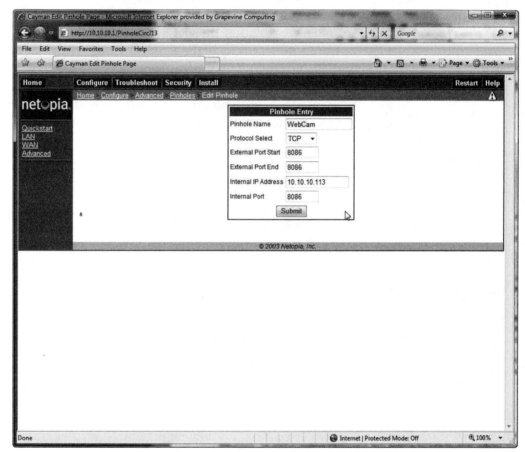

In my case I have configured webcamXP to use port 8086 instead of the default port 8080 it used. I then have to tell my firewall that all inbound traffic to a port I designate, again 8086, is to go to the IP address I use for my webcamXP server—10.10.10.113. Save these values and then restart your router/firewall for the settings to take effect.

You may have to experiment with the port number settings, because many ISPs block a variety of TCP/IP ports from being used to connect to subscriber computers and programs. To verify that your webcam server can be contacted by others on the Internet, call a friend and have them try to access your camera from their computer. You may have to use the trial-and-error method of figuring out which ports work and which ones do not, or contact your ISP for this information.

Looking back at Figure 6-4, in the upper-left corner of the page under Internal HTTP Server, you will see a value for Internet IP address. This is the Internet IP address of your cable or DSL modem and router during the time you are using your computers.

If your connection uses a dynamic rather than static IP address and your cable or DSL connection drops at any time, this IP address will change the next time the connection is reestablished. This means your friends will have to use a different IP address to view your webcam web server. I'll cover how to give your home network an Internet domain name to use in Project 15.

Feel free to explore the other features of webcamXP, such as the motion detector, which senses movement within the view of one or more cameras and can send an e-mail, record an entire video clip of the activity in range of the camera, and takes successive snapshots of activity the camera "sees"—a very useful security tool. If your camera supports pan and zoom features, these can be remotely controlled through the web server interface, and you can set up user lists with passwords so that your webcam server views are available only to those you select.

Project 7

Cellular Internet Connection to the Rescue

What You'll Need

- Your computer
- A cell phone with "tethering" capability and cellular data service
- A USB data cable to connect between your phone and computer
- Cost: Cable $10–50; Cellular data services $11–70/month

Whether you travel a little, a lot, or not much at all, many of us find ourselves in locations with no cable or DSL broadband Internet, no community or subscription Wi-Fi service, and only lousy dial-up—yet our cell phone works great. In these cases, rural, suburban or urban, it could be that cellular data technology is a possibility. Do you think...maybe...we could get Internet connectivity on our computer via our cell phone? At speeds higher than dial-up? Generally, the answer is yes.

In many areas, especially outer suburbs and rural areas, cellular data connections may be the only alternative you have to using slow dial-up or driving into town to visit the library or coffee shop. Admittedly, cellular Internet access is not for everyone—it's not cheap, it's not 6, 3, or even always 1.5 megabits-per-second (Mbps) broadband like the big cities have, but it is a lot faster than dial-up when you can get it. The cellular service providers have not achieved 100 percent phone coverage, much less data coverage, in the United States, but the coverage increases every month. For the cellular carriers, the top 10, 20, 50, and 100 urban markets are their first priority, but some are trying to get to the top 200 and top 500 most-populous areas faster than others.

While not limited or specific to home networking, we'll start with a modest project to get one computer connected by cell service, and expand from there. We're going to create an Internet connection configuration that looks like Figure 7-1.

Before we start, you'll have to work with your cellular carrier to make sure you have a phone, a data service plan that includes "tethering," the right USB data cable, and a CD-ROM (or downloaded files) that includes the appropriate drivers and

Figure 7-1

A data-capable cell phone can quickly become your PC's access to the Internet.

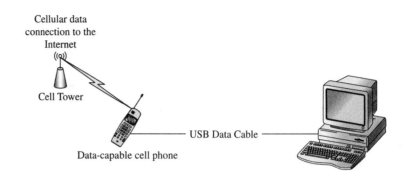

software to install on your computer so that your computer recognizes your phone as a USB or other high-speed modem. The cell providers should be more than happy to help—data service plans are pretty good revenue for them. From there we can get you going.

> **note** *"Tethering" is a term used by cell phone service providers to describe using your phone as a data modem, tethered by a special data cable, or through a Bluetooth pairing to transfer TCP/IP data between the cellular service towers to your computer through your phone. When tethering for data you cannot make voice calls on your phone.*

Step 1: Prepare Your Computer With Cellular Modem Drivers

Probably the toughest step of all is getting your computer to properly detect your cell phone as a USB-connected modem. USB devices and drivers can be the source of a lot of Plug and Play issues. While this process should be straightforward, there are a few critical things to attend to here:

1. Use the data connection software provided by your cellular service. For example, for Verizon, this software is called VZAccess Manager, and for AT&T, it is called Communication Manager. Different versions of each software package may or may not have the correct drivers for your phone. Sprint Nextel has its own branded connection software.

2. Make sure you have absolutely the right drivers (on CD-ROM or downloaded from the cell phone carrier or manufacturer) and install those drivers *before* you connect your phone to your computer. Reboot your computer if instructed to do so, but keep that CD-ROM or driver file handy.

3. Enable the USB modem link or USB-tethered feature of your cell phone. Some phones require that you do this every time you connect the phone to be used as a modem—otherwise, your computer may think it's just a mobile device and start synchronizing your address book. Without the proper

USB modem/data link setting in the phone, your computer will not properly sense the "modem," may become quite confused about the connection you made, and put the phone connection into "unknown device" lockout. (You may have to verify this with your phone/cellular provider when you set up your data service plan.)

4. When you do connect your phone to your computer, the computer's USB Plug and Play function should associate the driver you installed with the phone and install the driver automatically. For some devices, you may have to provide the driver CD-ROM or file to complete the process.

Step 2: Making Your First Cellular Data Connection

If everything in Step 1 works out correctly, your computer should be ready to connect to the Internet through your cell phone. For illustration I use a Verizon cellular data PC card and VZAccess Manager. The VZAccess Manager installation program uses a step-through wizard process, beginning with the option to check whether you have the latest version of the software, as shown in Figure 7-2. Typically, you will receive the latest or most appropriate copy of the software with your cellular data card and thus you do not need to perform the update check—which of course cannot be performed if you currently have no Internet connection. Click the Next button to continue.

Figure 7-2

The VZAccess Manager Setup Wizard is the entry point for discovering and configuring your cell-modem connection to the Internet.

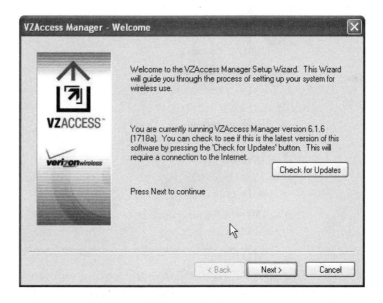

The wizard process then gives you the option to let the VZAccess Manager program be your single interface for Wi-Fi/802.11 and cellular/WWAN (Wireless Wide Area Network) connections, as shown in Figure 7-3. Most people prefer to let Windows or a separate Wi-Fi program handle this, so that is the option use in this project. Choose the Detect WWAN Device Only [1xEVDO/1xRTT/CDMA] radio button then click the Next button.

Figure 7-3

VZAccess Manager is best used to manage only your WWAN cellular connection to the Internet.

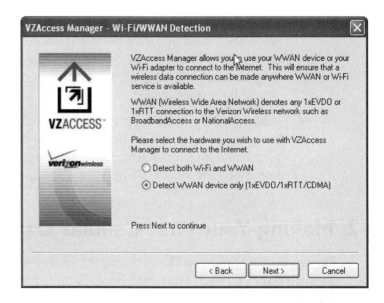

In the next, somewhat unnecessary step, the wizard pauses to tell you that it is going to try to detect your cellular data connection, as shown in Figure 7-4. Click the Next button to continue.

Figure 7-4

The VZAccess Manager wizard is the only way to detect and select the correct connection to your tethered or PC Card cellular modem.

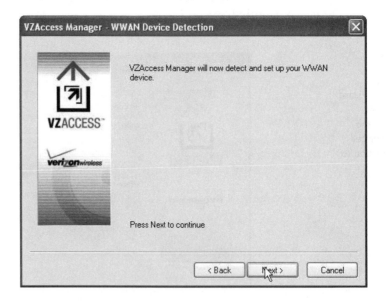

Since the VZAccess Manager program will work with many types of cellular data connections, you have to give it a head start by choosing the type of connection you have, as shown in Figure 7-5. The Data Cable option refers to PC-to-cell phone connection through a USB cable, and this connection type requires some reconfiguration of your phone to enable this option (consult your cell phone dealer). The PC Card option is the most common type of connection, while some phones support Bluetooth data modem operation. For our project, choose the radio button for Data Cable and then click the Next button.

Figure 7-5

Choosing the Data
Cable option begins the
correct cellular phone/
modem discovery
process.

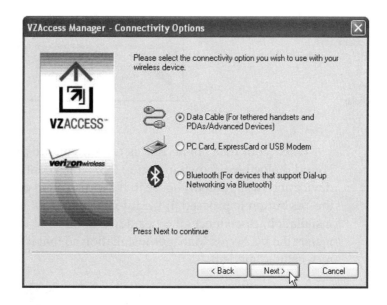

The wizard does not let you forget that it is important to ensure that the drivers
for your data cable-to-phone connection are installed. Though this warning, shown in
Figure 7-6, recommends that you cancel the VZAccess Manager installation process
to install the cable drivers, often you can simply connect the cable and your phone, let
Windows discover the new hardware, provide the drivers from a CD-ROM or downloaded file, and then continue with this installation by clicking the Next button.

Figure 7-6

The wizard's precaution
about proper drivers
for your USB data cable
and phone/modem
reinforces the need
to do drivers first,
connections second.

You will see a progress indicator (see Figure 7-7) as the installation process scans
every available communications port looking for a match to your cellular data cable
connection. Be patient as this can take 3–5 minutes to complete before the next dialog
box appears.

Figure 7-7

The USB cable/modem detection process is painfully slow but ultimately effective in detecting your cellular device.

When your data connection is found, you are presented with a Wireless Device Found dialog box, shown in Figure 7-8, which contains a description of the attached device and the option to choose it. If the displayed device information is correct, click the Yes button to proceed. If the information is not correct, you may have to stop the installation, disconnect all other devices, reinstall drivers for your phone, and then restart the VZAccess Manager installation so that it will detect only one, and the correct, cellular modem.

Figure 7-8

When discovered, the details of your cellular device and its virtual COM/serial port are displayed.

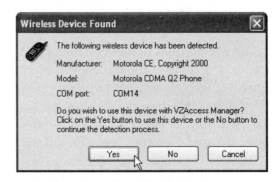

The wizard gives you a final confirmation option, shown in Figure 7-9, for the cellular connection you have chosen. Click the Next button to continue.

Figure 7-9

Confirm the wizard's device detection and you are a few steps away from Internet connectivity.

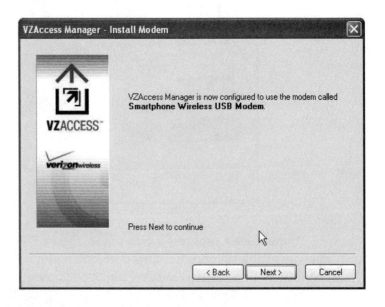

To ensure that your connection and data modem or cell phone are properly matched to make a connection, you must enter your cellular phone number (or the phone number assigned to the cellular data card), as shown in Figure 7-10. Click the Next button to continue.

Figure 7-10

Confirming the phone number of your cellular device allows it to be activated as a data device.

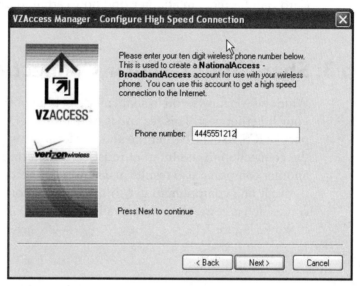

When the installation wizard is done, you are presented with the VZAccess Manager control panel, shown in Figure 7-11. You need to activate your modem with the cellular carrier by choosing Options | Activation from the menu bar. The activation takes about a minute, after which you can click the Connect button to go online.

Figure 7-11

The completion of the setup wizard presents you with a clean interface from which you can connect to or disconnect from your cellular data connection.

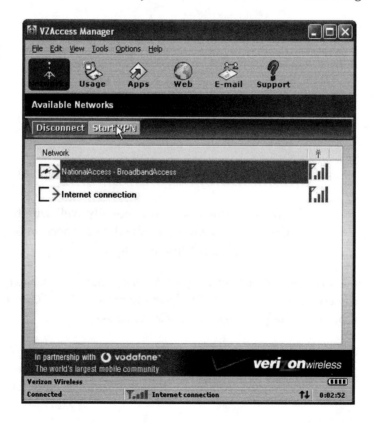

Verizon, AT&T, and Sprint try to make this process as painless and effective as possible (and generally succeed at doing so). Just like cable and DSL service, with cellular Internet connectivity established, you have a lot of options to take advantage of. As we move along through more projects, you will see how this works with a PC Card cellular modem adapter, and how certain phones and adapters fit into expanding your network just as cable and DSL modems do, as covered in previous projects.

Step 3: Share Your Cellular Connection

Windows XP allows you to share modem, Bluetooth, and wireless data connections to your Ethernet port. This feature is called Internet Connection Sharing (ICS.) A drawback of ICS, like the earlier file and printer sharing projects, is that the computer with the connection to the Internet must be turned on for this to work. Sharing through another computer also results in a slower connection—because the data has to travel through one computer to get to the Internet connection—but sometimes a slower connection is better than none at all or dial-up. Our expanded network schematic will look like Figure 7-12.

Figure 7-12

Using Windows XP Internet Connection Sharing to extend your cellular data connection to other computers through an Ethernet switch

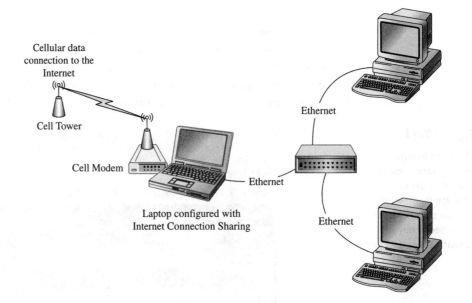

This ICS technique works equally well whether you are using your cell phone and USB data cable to make the data connection or you are using a PC Card cellular data modem plugged into a laptop.

note *PC Card adapters are not for laptops only. Though not a common item, you can find PC Card adapters that fit into desktop PCs so that you can use your PC Card devices (Wi-Fi or cellular adapters, for example) in a home or office computer.*

Setting up ICS in Windows XP brings with it Windows Firewall, so if you use a separate software firewall product (such as Norton Internet Security or ZoneLab's ZoneAlarm), you must disable Windows Firewall and then likely do some special configuration of your firewall software, or disable your firewall software and use Windows Firewall, which easily accommodates ICS. To set up ICS:

1. Right-click My Network Connection and choose Properties.

2. Locate and right-click the icon and label for your cell modem connection and choose Properties to open the Properties dialog box, shown in Figure 7-13.

Figure 7-13

Using Windows XP ICS to extend your cellular data connection to other computers through an Ethernet switch

3. Click the Advanced tab, shown in Figure 7-14, and then check all three of the check boxes under Internet Connection Sharing. Click the OK button twice to close these dialog boxes—ICS is setup and ready.

Figure 7-14

Enable ICS to work automatically for all computers on your local network.

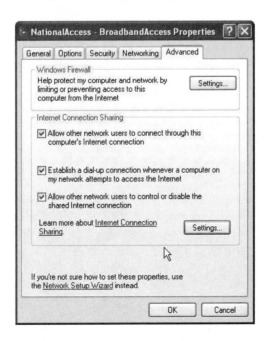

4. Connect all of your computers by Ethernet cable to an Ethernet switch. When Windows starts, your computers should acquire/assume a private network address to be able to communicate with each other, your ICS computer will sense requests to access the Internet, and your cell modem will connect—and then you're surfing the Web! Figure 7-15 shows a traceroute path from a computer sharing the cell connection, through an ICS computer, and finally to the CNET website address.

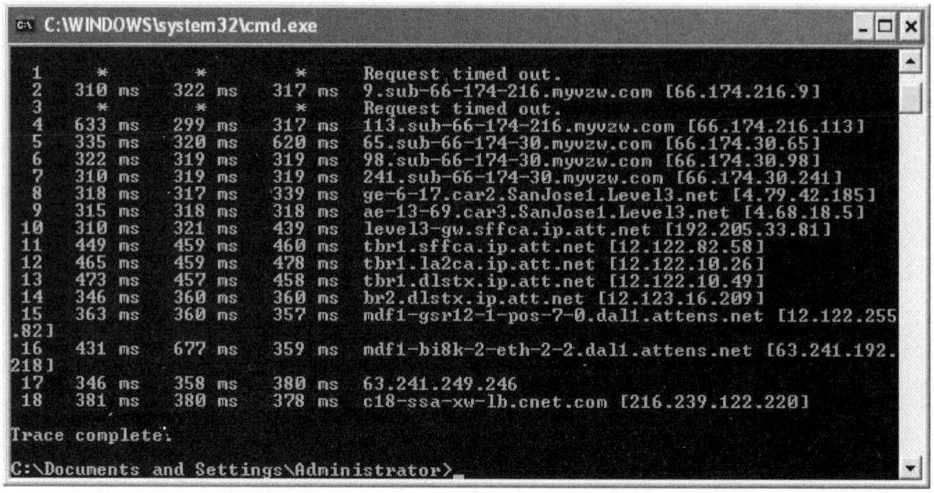

Figure 7-15

Tracing the route from a sharing computer, through the ICS computer, to CNET

```
C:\WINDOWS\system32\cmd.exe                                          _ □ x
  1     *         *         *       Request timed out.
  2   310 ms    322 ms    317 ms   9.sub-66-174-216.myvzw.com [66.174.216.9]
  3     *         *         *       Request timed out.
  4   633 ms    299 ms    317 ms   113.sub-66-174-216.myvzw.com [66.174.216.113]
  5   335 ms    320 ms    620 ms   65.sub-66-174-30.myvzw.com [66.174.30.65]
  6   322 ms    319 ms    319 ms   98.sub-66-174-30.myvzw.com [66.174.30.98]
  7   310 ms    319 ms    319 ms   241.sub-66-174-30.myvzw.com [66.174.30.241]
  8   318 ms    317 ms    339 ms   ge-6-17.car2.SanJose1.Level3.net [4.79.42.185]
  9   315 ms    318 ms    318 ms   ae-13-69.car3.SanJose1.Level3.net [4.68.18.5]
 10   310 ms    321 ms    439 ms   level3-gw.sffca.ip.att.net [192.205.33.81]
 11   449 ms    459 ms    460 ms   tbr1.sffca.ip.att.net [12.122.82.58]
 12   465 ms    459 ms    478 ms   tbr1.la2ca.ip.att.net [12.122.10.26]
 13   473 ms    457 ms    458 ms   tbr1.dlstx.ip.att.net [12.122.10.49]
 14   346 ms    360 ms    360 ms   br2.dlstx.ip.att.net [12.123.16.209]
 15   363 ms    360 ms    357 ms   mdf1-gsr12-1-pos-7-0.dal1.attens.net [12.122.255
.82]
 16   431 ms    677 ms    359 ms   mdf1-bi8k-2-eth-2-2.dal1.attens.net [63.241.192.
218]
 17   346 ms    358 ms    380 ms   63.241.249.246
 18   381 ms    380 ms    378 ms   c18-ssa-xw-lb.cnet.com [216.239.122.220]

Trace complete.

C:\Documents and Settings\Administrator>_
```

TiVo at Home

What You'll Need

- Your computer
- Your local area network
- USB-to-Ethernet network adapter
- TiVo Type 2
- TiVo Desktop Software 2.4
- Cost: USB LAN adapter $30; TiVo unit $200; TiVo Desktop Software, free

While TiVo told us it would change the way you watch TV, it now means a lot more than sitting in the recliner with a remote, time-traveling through show after show and skipping commercials. TiVo has changed a bit in the past few years—improved really—allowing you to do more across your home network and over the Internet.

The first steps to enjoying TiVo amidst your home network is getting it onto the network, which will allow you to make programming selections over the Web and get its content to your PC over the network. Let's TiVo...

Step 1: Connect TiVo to Your Home Network

The instructions provided by TiVo, http://tivo2.instancy.com/TiVoPDF/HMF_Guide_24.PDF, are fabulous. Just about any USB-to-Ethernet adapter will work for the physical connection—I've used a D-Link DUB-E100, and currently use a LinkSys USB200M, shown in Figure 8-1.

Once you have made the physical connection, the next step is to preset the network parameters for your TiVo DVR so that you can have precise control over network connections—especially through software and hardware firewalls.

When viewing your TV screen, press the TiVo button and navigate through Messages and Settings | Settings | Phone & Network | Change Network Settings menu items. Follow the menu and data entry selections to set a static IP address, gateway,

Figure 8-1

Connecting a TiVo to a network is as simple as plugging in a USB connection and Ethernet cable.

and DNS addresses for your TiVo DVR. These settings must follow the local network settings for your computers to allow computer-to-TiVo communications, and you must have a known IP address on your TiVo DVR to use if configuring communication to your TiVo DVR from the Internet. After you have finished configuring the new settings, restart your TiVo DVR.

Step 2: Communicate with Your TiVo on Your Computer

The TiVo Desktop software provides two very basic but not insignificant functions—it allows you to download from your TiVo and watch recorded shows through suitable multimedia software (Windows Media Player or Apple QuickTime), and it allows you to "publish" photos, music, and video from your computer to be viewed on your TiVo. The added bonus of being able to watch recorded shows with Media Player is also being able to transfer those recordings to a supported Windows media player. You can start using your TiVo DVR as a multimedia display server in about 10–15 minutes.

Download the TiVo Desktop software from www.tivo.com/desktop. Install the software, provide the Media Access Key from your TiVo DVR, and you are ready to select and transfer any or all TiVo-stored programs to your PC, as shown in Figure 8-2.

You need plenty of free disk space on your computer because an hour of TiVo content at High Quality makes TiVo files about 1.6GB in size. Each 1.6GB file takes

Figure 8-2

Transferring
TiVo-recorded
programs to your
computer

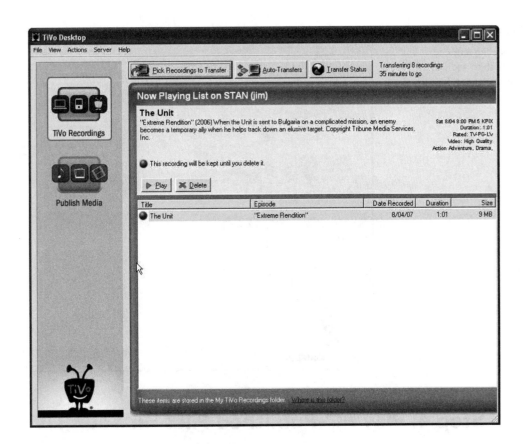

about 40 minutes to transfer across a typical home network. Select a recording to play, or in Media Player create a playlist or sync an episode to your Windows-compatible portable media player.

Step 3: Publish Computer-Based Files to Play on Your TiVo

Your TiVo can play music and videos as well as display pictures you share from your TiVo Desktop–equipped PCs. This handy little feature sure beats dragging a PC into the family room to have the relatives huddle around a 17″ screen to view your vacation pictures, or jacking in to your stereo to play MP3s.

To let your TiVo solidify its place as home entertainment central, install the TiVo Desktop software on any and all PCs that contain media files you want to share and play on your TiVo-connected TV and stereo.

Click the Publish Media button on the TiVo Desktop software, shown on the left in Figure 8-3, and then click either the Music, Photos, or Video tab.

On the Music tab, for example, click the Add Music button. This presents a file browser dialog box in which you can select a folder and music files to publish. Once published, files are ready to be played on your TiVo.

Figure 8-3

Selecting music to
publish to and play
through your TiVo

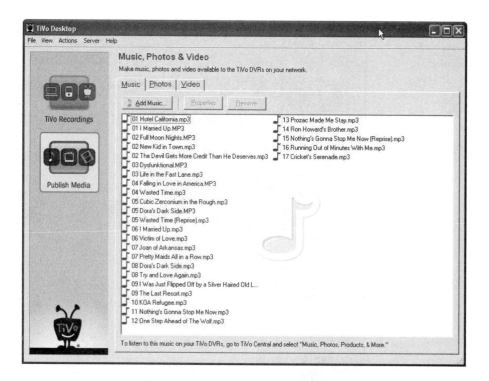

Step 4: Play PC-Published Files on Your TiVo

At your TiVo, press the TiVo button, select Music, Photos, Products, & More (see Figure 8-4), wait a few seconds, and then scroll to select the computer with published media, as shown in the example in Figure 8-5.

Figure 8-4

The familiar TiVo
Central menu is the
portal to access viewing
options and settings.

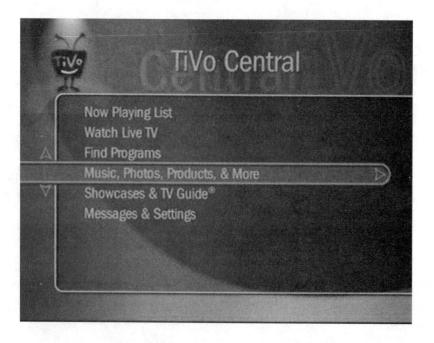

Figure 8-5

Select the publishing computer, pick a song to listen to or a photo to view, and start enjoying your new multimedia experience.

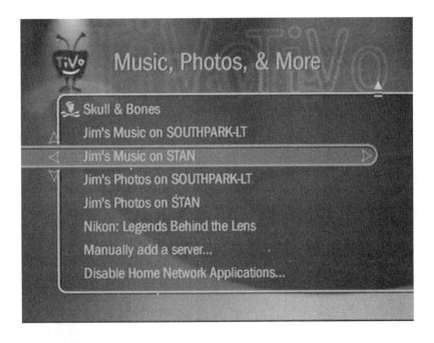

Software Firewall Settings for TiVo Desktop

There are circumstances where the TiVo Desktop software and your TiVo simply will not communicate—especially if you have a software firewall (such as ZoneAlarm, Black Ice, Norton Internet Security, or McAfee Personal Firewall). Normally these products allow communications between software you intentionally install and other hosts on your network, if not, check and reconfigure the firewall settings to allow the following ports access between your computers and TiVo so that the TiVo Desktop server can communicate with your TiVo on your local network:

- TCP port 37
- TCP port 2190
- TCP port 4430
- TCP port 7287
- TCP port 7288
- TCP port 8000
- TCP ports 8080–8089
- UDP port 123
- UDP port 443
- UDP port 2190

If you want to set up specific computer-to-TiVo access connections/routes, configure static IP addresses on your computer(s) and TiVo.

Project 9
Sling Media Bridge

What You'll Need

- Your computers
- Your home network
- Slingbox AV unit
- SlingPlayer
- Cost: Sling AV $149

This project could keep you from ever buying or replacing a TV set again, and could make you ignore your present set or toss it out. We're talking about moving your TV viewing from a bulky box or an expensive flat-panel TV to your PC screens—heck, you've got PCs all over the place and you spend all of your time with them, so you may as well add a little TV to your screen.

The catalyst and means to take this drastic measure is the Slingbox AV appliance, shown in Figure 9-1. Sling Media's Slingbox AV adapts video content, S-Video or composite video, and accompanying audio to a broadcast stream over Ethernet. The stream is viewed through a SlingPlayer application, with versions available for PC and Mac. A computer with the SlingPlayer may be on your home network, for best performance, but may connect to and view streams from the Slingbox anywhere on the Internet.

Step 1: Install the Slingbox AV

To install the Slingbox AV, four connections are required—power for the Slingbox, an Ethernet cable to your network, your choice of S-Video or composite video and audio to catch video programming, and an IR remote cable to relay channel changing commands from your PC to your TV or DVR, as shown in Figure 9-2. The video source can be a cable or satellite receiver, DVR, DVD, or tape player—if it outputs S-Video or composite video, it is an eligible source. The IR cable is used to place a remote control transmitter element near the remote receiver sensor on your video source.

Figure 9-1

The Slingbox AV is a simple and powerful adapter to broadcast video over a network.

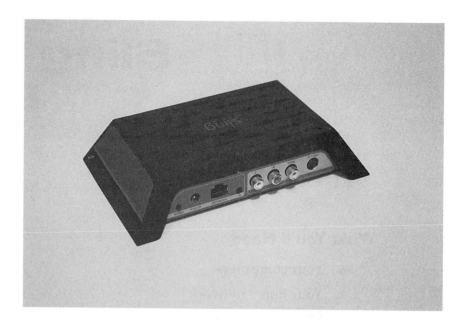

Once connected to your network, the Slingbox AV acquires an IP address through the DHCP services of your router, and becomes discoverable by the SlingPlayer software.

Figure 9-2

The Slingbox AV uses typical audio and video connections—here to a TiVo Series 2.

Step 2: Install and Configure SlingPlayer

Installation of the SlingPlayer software takes you through a dozen screens to help you identify your Slingbox AV unit (it also works with the Slingbox PRO, Slingbox Tuner, and Slingbox Classic), assign a password, select the video source (see Figure 9-3), test the remote control connection, and verify video quality and network connections. After that, you are good to go.

Figure 9-3

Slingbox AV supports
a tremendous variety
of video sources, from
cable and satellite
receivers to DVRs.

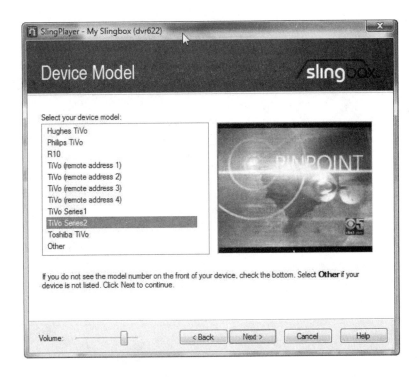

If you want to make your Slingbox AV viewable by other SlingPlayers over the Internet, the setup process can also help you configure your router/firewall, as shown in Figure 9-4.

Figure 9-4

Verify or configure your
Slingbox AV's network
settings, including
which port it will accept
Internet connections
from.

Install the SlingPlayer on any computer you want to watch your video source on, and you'll have TV, even TiVo, anywhere you roam on your network or, if you like, across the Internet.

Part II

Advanced

Sharing Your Basic DSL Broadband Connection

What You'll Need

- Setup from Project 1
- A second Ethernet network cable
- Ethernet router
- Cost: Router $40–80

After you complete this project, you'll be able to share your broadband connection among four computers, and gain some recommended protection against hackers. Adding a combination firewall/router alters the connections of your network and adds some critical configuration details we'll be covering.

The schematic of your basic broadband connection looks like Figure 10-1; your DSL (or cable) modem connects directly to your computer—pretty simple.

To this we'll add a small network router (see Figure 10-2), which takes the connection from your broadband modem, processes it, adds protection to it (more on this later), and then distributes the network data to up to four computers. The connections to the router include power, wide area network (WAN) for the Internet connection to be shared, and local area network (LAN) for the computers connecting to it.

Figure 10-1

Typical basic network connection—a broadband modem and computer.

DSL Modem   Computer

Figure 10-2

A common broadband
network router

Step 1: Connect the Router

This step is straightforward—disconnect the network cable from the broadband
modem. Connect the network cable to one of the four LAN ports on the back of the
router, shown in Figure 10-3.

Figure 10-3

Rear view of broadband
router showing four
LAN, one WAN, and
power connections

Connect the power adapter and turn on the power to the router. It will take a
minute or two for the router to start up; meanwhile, turn on or restart your computer.
When the computer is ready to use, it will have established a connection with the
router and obtained an IP address. At this point you can configure your router for
your broadband connection.

Step 2: Check Your New Network Connection

The directions for your router should tell you the connection sequence and how to determine that the connection with your computer is working correctly. We will verify that it is working by checking for the IP address your computer should obtain from the router. To do so, we will use the examples of both the Windows Status dialog box for your network adapter and a DOS command under Windows—IPCONFIG. Hopefully you won't have to use IPCONFIG more than a few times, but if and when you do, this small program can tell you a lot about the Ethernet TCP/IP connection.

note *The Windows IPCONFIG program is a powerful utility that can show you some very basic to fairly detailed information about your computer's network connections - the most basic being your computer's IP address and the IP address of the gateway leading from your network to another network or the Internet. Using the command-line IPCONFIG/all will reveal significantly more details about your Ethernet, wireless and other types of network connections. An equivalent command in Unix (and the Mac X OS) would be "ifconfig."*

Assuming you've got a Windows PC, access the network status as follows:

1. Click Start.

2. Right-click Network Places.

3. Choose Properties.

You should see the icons for one or more network connections, and a status of Connected, as shown on the right in Figure 10-4. If you do not see a status of Connected, check the network cabling and directions for your router to correct the situation.

Figure 10-4

Windows Network Connections window showing the status of network adapters

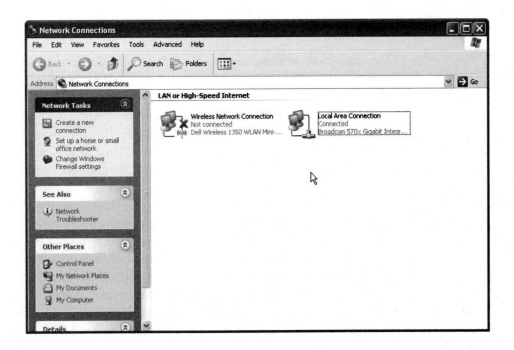

4. Next, right-click the Connected network adapter and choose Status. When the Status dialog box appears, click the Support tab. You should see details similar to those shown in Figure 10-5.

Figure 10-5

Network status showing the IP and gateway address for the network connection

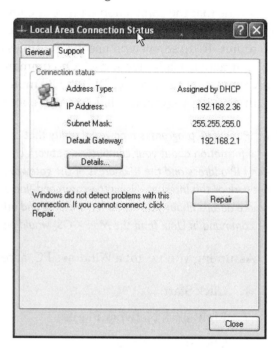

Note the IP address for the Default Gateway—this is the IP address you will use in your web browser to access the configuration software for the router. (This address should also match the documentation for your router.)

Step 3: Configure Your Router

Open your web browser, and in the Address field, enter the Default Gateway IP address from Step 2. Browsing to this address should bring you to the router's configuration home page, as shown in the example in Figure 10-6.

From the router's web interface, you have complete control over your Internet connection—from making and disconnecting the connection, to allowing desired content in, to filtering out undesired content. Your router can "dial in" and authenticate your DSL connection as needed or keep the connection alive all the time.

The first task here is to configure the router for the type of authentication your broadband connection uses. Most DSL connections use the Point-to-Point Protocol over Ethernet (PPPoE), which requires a username and password to log onto your ISP and complete the connection—a security measure many DSL providers use to help ensure the control over and safety of your connection.

On the next screen (see Figure 10-7), you fill in your ISP account information. Your account username is typically your full e-mail address. The Service Name and MTU items usually do not need any changes. The disconnect time period is used only if you want your DSL connection to drop if it is not being used. This feature saves

Figure 10-6

Belkin router
web configuration
page showing DSL
connection choices

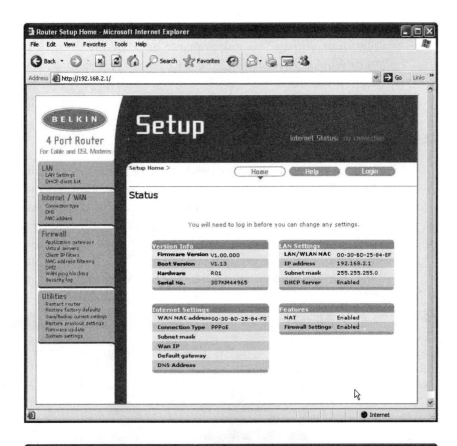

Figure 10-7

DSL service logon
account configuration
with connection
timeout feature

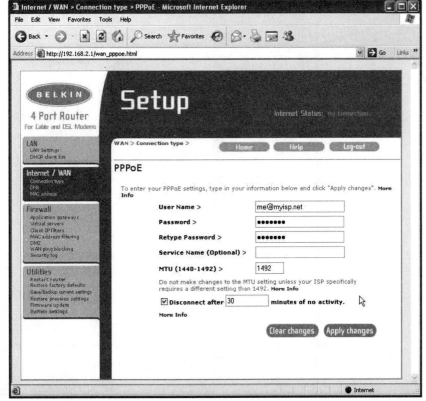

your ISP some bandwidth, and if your modem and PCs are disconnected from the Internet, you are less vulnerable to hackers. If you run a server such as a web camera or connect to your home PC by remote control, you probably want to disable the timeout feature.

Step 4: Secure Your Router

Almost every router provides some level of security against accidental or intentional reconfiguration of the settings you make. Some routers require only a password (typically "admin" by default), some a username and password (usually "admin" and "password" by default to start with), while others require no password initially but allow you to set one. It is strongly recommended that you choose at least your own secret password to prevent tampering. This setting is configured on a system or security configuration menu, as shown for this router in Figure 10-8.

Figure 10-8

Belkin router internal System Settings page for router password, date, time, and remote management

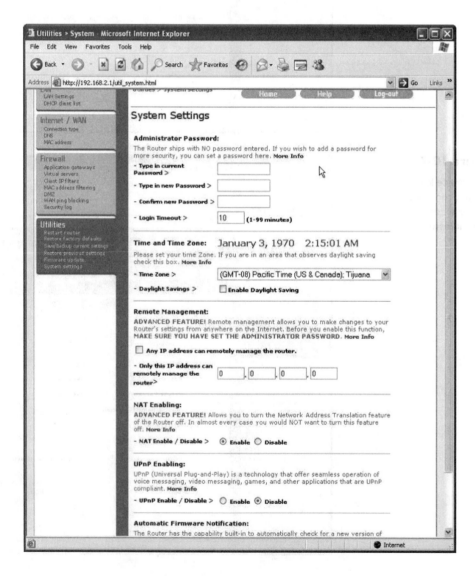

You may also have the option to enable or disable remote management of the router. If you enable it, anyone on the Internet may be able to tap into your router and change the settings, so you do not want this feature enabled unless or until you need help from your ISP or other support resource (then disable remote management when the support session is completed). The date and time setting is used to place an accurate timestamp on internal logging that records possible attacks coming from the Internet or unusual activity inside your home network.

Completing the preceding steps makes your router ready to be connected to your DSL modem, and then the Internet, and ready for you to share the connection with up to four other PCs—or more, depending on the configuration of the rest of your home network or the addition of a wireless network access point. The final step is connecting the router's WAN port to your DSL modem.

Step 5: Connect Your Router to Your Modem

Of the six connections on your router, the WAN port is the one that connects to your DSL (or cable) modem. Connections beyond your own LAN to other locations, in this case the Internet, are considered a wide area network. This is the port to and from which Internet traffic flows and is filtered for your local computers. All network activity on the LAN side of the router stays on the LAN unless traffic must flow to other locations—the Internet at large or perhaps a connection to your office.

Connect an Ethernet cable from the WAN port of your router to the LAN or Ethernet port of your DSL modem. When the connection is made, your router controls and distributes Internet data to your computers.

The completed wiring of computer, router, and DSL modem is shown in Figure 10-9. You may change the cable to be of any convenient length up to 300 feet—

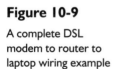

Figure 10-9

A complete DSL modem to router to laptop wiring example

so one cable could connect to a computer close to the router, and others could connect to computers across the room or in other rooms, or to a wireless access point. The schematic of your new network configuration is shown in Figure 10-10.

Figure 10-10

Expanded network schematic showing the addition of the router and connection of up to four computers

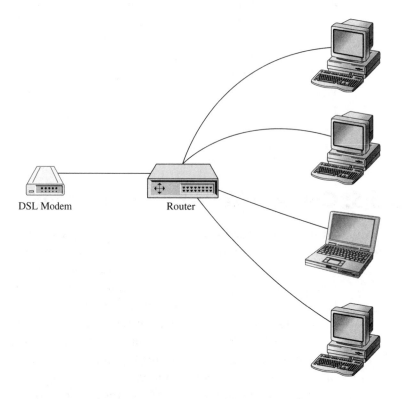

DSL Modem   Router

With this enhanced configuration, your network is ready for just about anything. The projects ahead will exploit both Internet and internal network sharing capabilities, and help you avoid others exploiting your network from inside and out.

Configuring
Firewall
Protection for
Your Network

What You'll Need

- Setup from Project 10
- Your computer and web browser
- Cost: $0

With a broadband connection and a combination firewall/router, you have all you need to really enjoy the Internet and keep yourself from being hacked—that is, anything from having your personal information stolen, your computer abused as an unwitting virus- or spam-spreading robot, or simply infected and rendered nearly useless.

Think of a firewall as an Internet doorman—what you want to view across the Internet is openly available to you, and you may come and go as you please. Anyone or anything on the Internet that wants or needs to get to your computer is hopefully stopped cold at the doorway—allowed to get in only if you give "the nod" to the doorman.

Firewalls work in mysteriously wonderful ways, detecting potentially bad traffic trying to get into your network, with some even blocking known bad traffic trying to get out. Sometimes a firewall can work too well for what you want to do. For example, a firewall might block Internet phone calls that you want to accept, reject remote control from people you ask to help support you with computer problems, stop you from accessing your office network through VPN, or prevent you from using a webcam. Success with a firewall depends on what data traffic you allow to pass through your firewall to and from the Internet, with a few basic examples.

This project will take you through the steps of making your internal network safe from if not completely invisible to the Internet at-large, and allowing only certain types of application data (for instance, remote control of computers on your network or running your own web server) into your local network from the Internet.

Step 1: Find Your Router's Firewall Settings

Access your router's configuration web pages by typing your router's IP address into your browser's Address bar—in our example from Project 10, the router IP address is 192.168.2.1. Among the available menu items should be a selection to configure the router, or directly access the firewall settings. In our case the firewall selections are clearly visible along the left menu. Clicking Firewall takes you to a basic configuration selection—whether or not you want to enable or disable the firewall—as shown in Figure 11-1. Normally you want the firewall enabled, but you may have to disable it if you're working with a support technician to troubleshoot problems in your network or Internet connections.

Figure 11-1

Controlling whether your router's firewall is enabled

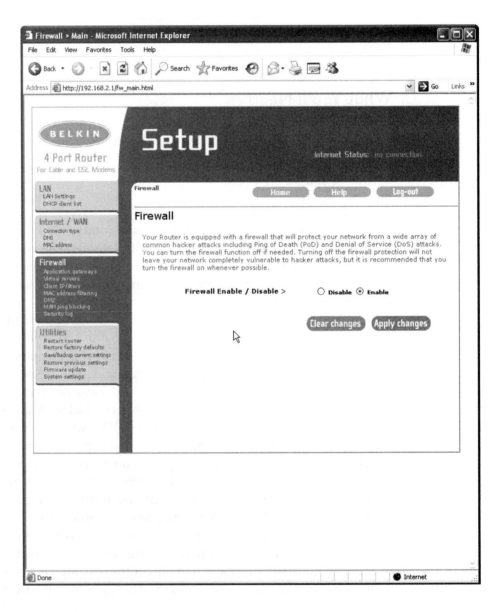

With the firewall feature disabled, all of the computers on your internal network are vulnerable to almost anything and everything floating around the Internet. You're normally safe from and immune to the e-mail, web browsing, music downloading, and normal traffic because that network traffic is well defined and flows between intended sources and destinations just fine—your ISP's routers and hundreds of others along the Internet keep it away from you.

Your ISP may tell you that it filters a lot of bad Internet traffic and that you are safe using its services, but in reality, if your ISP filtered the Internet, you would not have a real, whole Internet connection or experience. Many ISPs block some Internet services—to prevent or at least make it more difficult for you to host web sites, mail servers, or file servers on your home connection—but there are many ways around those restrictions, and none of those restrictions can do much to prevent abusive, exploitive "robots" from wandering around loose looking for vulnerable places to land.

The threats you face are hundreds if not thousands of tiny programs running on others' computers that do nothing but poke and prod and probe looking for soft spots such as unprotected computers and networks. When these robots find a soft spot, they may not exploit it directly—instead, they may report back to a master robot, which then commands other robots to infiltrate potential victims. If you appreciate science-fiction drama, you'll appreciate the actions of the Internet underworld.

If this is getting a bit scary, let's tone it down a bit by enabling that firewall feature in the router and, for good measure, the firewall of Windows or add-on product such as ZoneLabs ZoneAlarm or Norton Internet Security. Your router's firewall, also known as a hardware firewall, is basically a software program in the router that inspects data coming into your Internet connection. If the firewall software finds something known as bad data, or even suspected as bad or rather abusive data, it should stop it cold. Your computers will never even know the bad stuff is out there.

Step 2: Hide Your Connection

Enabling your firewall is the first part of providing a protective layer to cover your Internet connection. The next part is to set a feature that keeps casual snoopers from finding your connection. This feature blocks the basic ping that alerts a snooper that there is an active connection available to exploit.

Keeping simple pings from finding your connection is usually controlled by a setting referred to as WAN ping blocking. This setting is easily found on the Belkin router's configuration menu, as shown in Figure 11-2. Enable this setting unless your ISP, connection provider, or other support tech needs to verify your connection.

Figure 11-2

Preventing your router
from responding to
basic ping inquiries

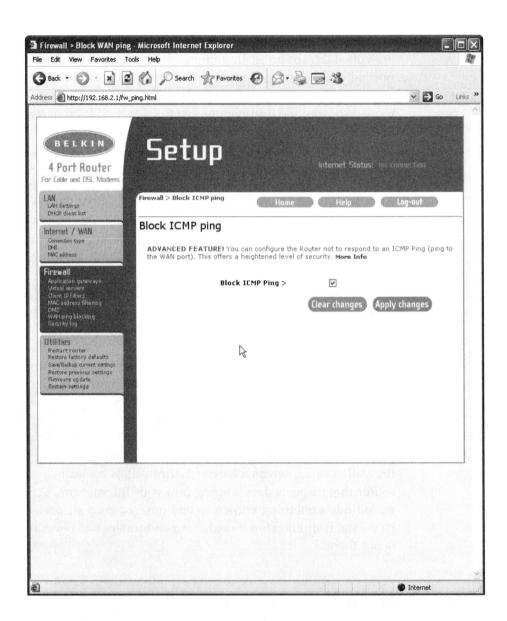

Step 3: Allow Applications Through Your Firewall

Almost everything coming from the Internet is blocked, excluding e-mail and web content and file downloads you requested to come to your computers. This blocking can prevent inbound Internet phone calls, viewing your webcam, playing games, and other common activities we are promised we can enjoy over broadband connections. Sometimes you want to let Internet traffic onto your network.

Fortunately, most home firewall/routers can be set up to allow specific programs and data to come into your network, to get to your computers. There are dozens of common, everyday, run-of-the-mill games and special programs that you or your family may want to enjoy with your friends and family elsewhere across the Internet that will not work until you reconfigure your firewall to allow them to work. In many ways, as you try to do more with your networked computers, you are discovering how and why the networking and Internet does or does not work as advertised—sometimes you have to tell your computers and network equipment exactly what you will or will not allow them to do or who they communicate with.

Many of us are starting to explore Internet-phone or Voice over IP (VoIP) services as a way to avoid or reduce the costs of long-distance and overseas phone calls. Skype is one of the better-known Internet-phone applications, but Yahoo!, AOL, and Microsoft instant messaging services also provide a variety of user-to-user voice and video options. For the full features of these applications to work, whether you use a Windows PC, an Apple Mac, or a Linux computer, you need to instruct your firewall to allow data specific to these applications to flow freely from the Internet into your computer.

Access your firewall's configuration page, which may contain references to various applications, specific ports and protocols, or both. In our example router/firewall, there are two pages, shown in Figures 11-3 and 11-4, both of which have

Figure 11-3

Opening your firewall for specific applications

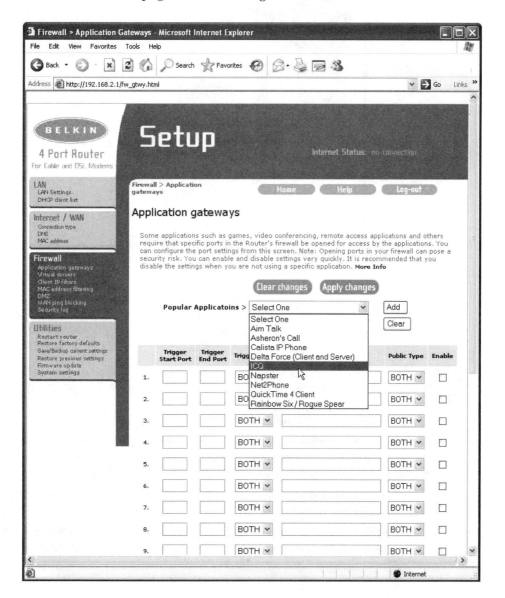

Figure 11-4

Configuring your
firewall for custom
applications and ports

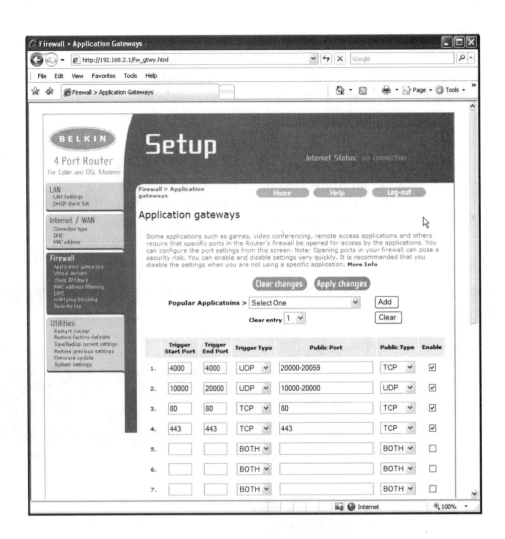

settings for specific applications and custom settings. Look up your application or follow the directions provided with it to allow UDP, TCP, or both data protocols into your network on the specific port or ports listed for them.

Step 4: Configure Your Firewall for Internal Servers

In addition to allowing Internet traffic for specific applications onto your network, you may want to allow someone outside your local network to access only specific computers inside your local network , such as an FTP server or web server, to do only specific tasks, such as remotely control your computer, or access a web camera. Figure 11-5 shows the router configuration screen for virtual servers. Though technically not an accurate description in this case, this page tells your router that if an outside computer wants access to a web server, wants remote computer control, or wants some other application on your local computers, your router should direct that request to a specific computer, and only that specific computer. These configurations are typically called "pinholes"—a very restrictive peek from outside to inside.

Many routers offer preconfigured pinholes—such as one for pcAnywhere, a remote-control program that lets someone "drive" the keyboard and mouse of a computer and see what is happening on the screen, which is very useful for technical support. Similar preconfigured pinholes may exist for Internet phone, or you can configure customized pinholes for webcams, radio communications services, and almost anything else you can imagine.

Figure 11-5

Configuring a pinhole for a specific application to a specific computer inside your network

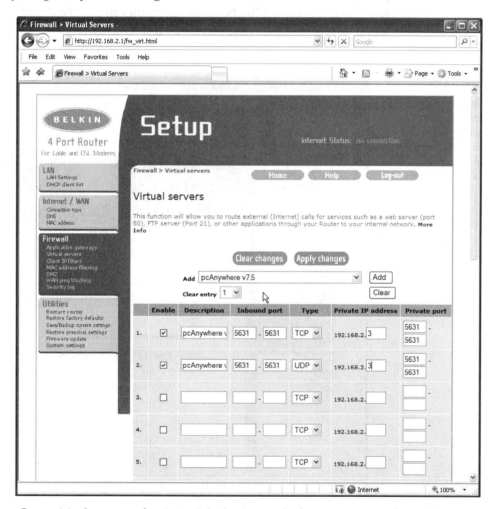

One critical aspect of using pinholes instead of opening your firewall to general application use that lets traffic flow to many computers (described earlier in the project, and typical for multiplayer games) is that for a pinhole, the request coming from outside will be directed to one and only one computer on the inside.

For a pinhole to be configured and work, you must know the IP address of the specific computer that you want to allow the data to get to, and you must set a fixed IP address on that computer instead of using dynamic addressing. You must also know the TCP or UDP data ports the pinhole will move data through. Applications that require pinholes typically document all of these factors for you—all you need to do is type them in.

Step 5: Allow Unfiltered Access to One Computer

There may be a rare occasion when you want to expose one computer to the whole unfiltered, unprotected Internet without exposing your entire local network—this is called creating a "demilitarized zone," or DMZ.

Locate the DMZ option in your router's configuration options (see Figure 11-6), and then specify the IP address of the one device or computer that will receive completely unfiltered, unprotected Internet connectivity. Beware—this is like drinking from an uncontrolled fire hose: the computer in the DMZ will have no protection unless it is built into the computer or by a software firewall.

Figure 11-6

Configuring a single IP address to be allowed completely unfiltered access to and from the Internet

Home Network Workgroup

What You'll Need

- Your computers
- Your local network
- Cost: $0

Having your PCs connected to the same network is the first step to a real home network. Sharing an Internet connection is a terrific use of a home network. As you might imagine, it is not the only benefit you can explore. File sharing (documents, media, photos) and printer sharing, from Projects 3 and 4, are two of the most common and valuable things you can do with your home network.

Before you can accomplish these feats, your computers need to know about each other—be part of the same computer-to-computer affiliation. Business networks use the Internet-centric technology of domains and domain name services to identify and communicate with each other. Home networks use workgroups and communications features of Microsoft networking technologies to know about and communicate with commonly grouped computers. A common workgroup or association among a collection of computers makes it possible for computers and easy for users to be aware of, identify, and communicate between each other. Creating just such a workgroup among computers is what we'll accomplish in this project.

Step 1: Establish a Common Workgroup

Establishing a common workgroup name among all of your computers makes it easier to locate other computers on your common network without knowing a lot about their IP addresses or computer names. Under Windows XP, this takes about six mouse clicks and a restart of your computer. Under Vista, you have to endure eight mouse clicks. The window at the heart of this operation is shown in Figure 12-1.

Figure 12-1

Descriptive names for your computer and workgroup help you easily identify and connect to each computer on your network.

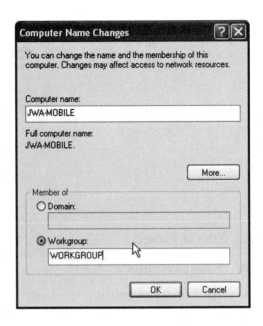

For XP:

1. Go to Start (lower left corner of the Windows desktop), right-click the My Computer icon and choose Properties.

2. Click the Computer Name tab and click the Change button.

3. In the Computer Name Changes dialog box, choose the Workgroup radio button and then type in a name of a grouping you want all of your computers to be part of.

4. Click the OK button. You'll get a pop-up message telling you that you must restart your computer for the changes you made to take effect. Click the OK button to close the System Properties dialog box and then restart your computer.

For Vista:

1. Go to Start (the Windows icon in the lower left corner of the desktop), right-click the My Computer icon and choose Properties.

2. In Control Panel double-click the System icon. In the left column click Advanced System Settings to access the System Properties dialog box.

3. At the top of the System Properties dialog box click the Computer Name tab.

4. Choose the Workgroup radio button and then type in a name of a grouping you want all of your computers to be part of.

5. Click the OK button. You'll get a pop-up message telling you that you must restart your computer for the changes you made to take effect. Click the OK button to close the System Properties dialog box and then restart your computer.

Repeat these instructions for all of the computers on your home network. When you are finished placing all of your computers into the same workgroup, all of the computers' names should appear in the window when you open My Network Places (XP) or Network (Vista) window.

note *If you bring home a computer from your office, it may be a member of a different workgroup or in an Active Directory domain, defined by your office computer support people. You can usually change the workgroup name to match your home network, but be careful to change it back before you connect at the office. You will probably not be able to change from Domain to Workgroup mode unless you have domain administrator rights on your office network (few of us should or do).*

Step 2: Verify NetBIOS Settings

If you cannot "see" one or more computers within your new workgroup, it is probably because one or more computers do not have Windows NetBIOS network communications enabled in the Advanced TCP/IP Settings dialog box, shown in Figure 12-2. It is enabled by default unless explicitly disabled. To check this, follow the instructions for your operating system, provided next.

Figure 12-2

NetBIOS is a network communications technology that allows computers to identify each other and be found on a local network.

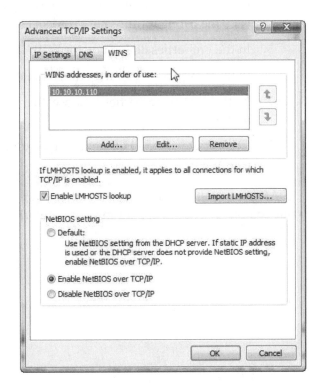

For XP:

1. Right-click My Network Places and choose Properties.

2. Locate and right-click your network adapter and choose Properties.

3. In the Properties dialog box for your network adapter, scroll to and double-click the listing for Internet Protocol (TCP/IP).

4. In the Properties dialog box for TCP/IP, click the Advanced button.

5. In the Advanced TCP/IP Settings dialog box, click the WINS tab.

6. In the WINS properties, choose the Enable NetBIOS over TCP/IP radio button.

7. Click OK to close all of the dialog boxes and windows and then restart your computer.

For Vista:

1. Right-click My Network places and choose Properties.

2. In the Control Panel double-click the Network and Sharing Center icon window, in the left column click Manage Network Connections.

3. Locate and right-click your network adapter and choose Properties.

4. In the Properties dialog box for your network adapter, scroll to and double-click the listing for Internet Protocol (TCP/IP).

5. In the Properties dialog box for TCP/IP, click the Advanced button.

6. In the Advanced TCP/IP Settings dialog box, click the WINS tab.

7. In the WINS properties, choose the Enable NetBIOS over TCP/IP radio button.

8. Click OK to close all of the dialog boxes and windows and then restart your computer.

After completing these steps on all of your computers and restarting them, all of your computers should appear in your network.

Creating Computer User Accounts

What You'll Need

- Your computers
- Your local network
- Cost: $0

A stand-alone personal computer that is accessible by only one person is usually pretty safe, secure, and private (Internet considerations aside) and thus doesn't usually warrant requiring a username and password to log on. In fact, the default option for most operating systems is to simply start up and present a main screen, ready for any command or type of work. By contrast, on a network, it is prudent to exercise a certain amount of per-user control for computers by creating user accounts.

User accounts in company settings provide minimal to extensive data privacy and accountability within documents, shared server resources, and the overall network. At home, user accounts serve a similar purpose—but at home, the lack of separate and secure user accounts is not likely to make you fail an FDA, SOX, or ISO-9001 information security audit. At home, you should be concerned about such issues as identity theft and loss of homework, family photos, bank and tax records, and other things that are important to you and your family that are stored on computers.

Every Microsoft and most recent Apple operating systems offer some type of user accountability, but not always security in separating and protecting the access to data on a per-user basis. Windows NT, 2000, XP, and Vista all offer this protection, and it comes in pretty handy at times.

Having separate user accounts forms the basis of determining how much control each user has over the resources on a computer, and to some extent how much damage or good they can do to the files and programs. User accounts typically take on one of three levels of control, privilege, and rights on a computer system:

- **User** A User-level user account can manage their own documents and files, but cannot manipulate the files of others, add or remove printers or other hardware, or install or remove programs.

- **Power User** A Power User–level user account can install printers, programs, and so forth but cannot manipulate other user accounts or do serious harm to the computer or file system.

- **Administrator** An Administrator-level user account can do everything—to some extent, even control other users' documents.

Best practices dictate that no one be an Administrator-level user for day-to-day work, but many of us are. Access to the Administrator level gives ultimate power to people, viruses, hackers, and malware. If you are concerned about security and must administer one or more computers, set up two accounts—one as User or Power User for day-to-day work, and another separate account as an Administrator-level user to control the system completely.

Establishing user accounts also allows you to control who gets to see, share, and work on documents of other users and on other computers when files and printers are shared, as described in the next two projects. A good way to start this process is on paper, or whiteboard—list all of the possible usernames, which computers they will use or share files and printers from, and how much control they need at each computer. With a strategy mapped out, your work goes pretty fast.

Step 1: Create User Accounts

Windows XP and Vista provide two ways to access user account management tools. If you use the Control Panel and wizards method, you can easily miss some attributes found in the more techie but direct means of the Computer Management console. My preference is to take you into this task through the Computer Management console because you will have to go there eventually, and it is much faster when dealing with multiple users and their different specific access rights. By the way—you have to be logged on with an account that has Administrator rights when managing user accounts.

1. Right-click My Computer and choose Manage.

2. In the left pane of the Computer Management console, shown in Figure 13-1, double-click Local Users and Groups, and then select Users. In the middle pane you will see a list of all known users for this computer.

3. Right-click either the Users item in the left pane or anywhere in the center pane and choose New User to open the New User dialog box, shown in Figure 13-2, where you begin adding a new user account to this computer.

Figure 13-1

Windows' Computer Management console provides access to many attributes of your computer, including the users and types of users available.

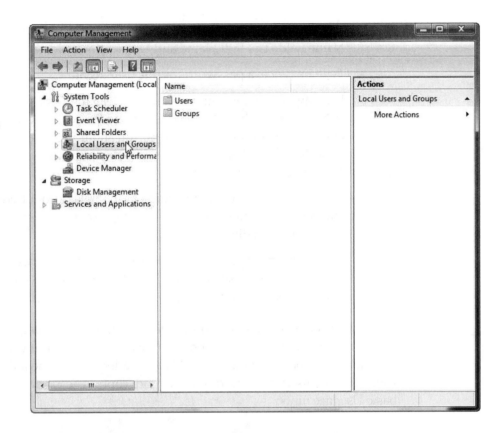

Figure 13-2

The New User dialog box provides the basic elements to establish a user's account on the local computer.

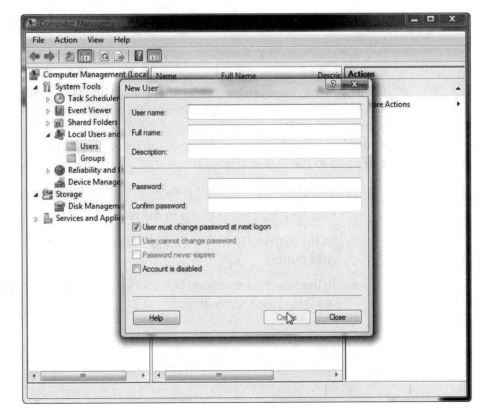

4. Type in the logon name you wish to use for the computer user in the User Name field. The Full Name and Description fields are optional.

5. Type in, twice, a password for this user account.

6. For a typical home setting, you can clear the User Must Change Password at Next Logon check box and check the Password Never Expires check box. Out of respect for the other users, you should tell them to change their password to one of their choice—as an Administrator, you can always reset the password if they forget it.

7. Click the Create button to complete the new account creation process. You will return to a fresh New User dialog box for the entry of more new user account information.

8. Repeat this process for all of the user accounts you need to create. After creating the last user account, click the Close button to return to the Computer Management console.

Each of the new accounts you create is set to User privilege—they cannot install or remove software or printers, nor do much of anything but use software and manage their own documents. This level is quite adequate for most day-to-day computer use, but some users may need to have more control—especially to help and support basic User accounts.

Step 2: Set Account Privileges for Each User

It is not uncommon for some users to need computer account privileges that allow them to install their own software or configure printers, but that do not give them full rights on the system. For such users, Microsoft established the Power User account level. You can manage account levels through the Member Of tab in the user account's Properties dialog box to add the user to one or more groups, or you can add multiple users to an account group in the respective group's Properties dialog box. To add users to the Power Users group:

1. Click Groups in the left pane of the Computer Management console.

2. Double-click Power Users in the center pane.

3. In the Power Users Properties dialog box, shown in Figure 13-3, click the Add button.

4. In the Select Users dialog box, shown in Figure 13-4, type in the users' names, separated by a semicolon.

Figure 13-3

The Power Users group provides an appropriate privilege level for many users—not too much control but not too many restrictions for daily use.

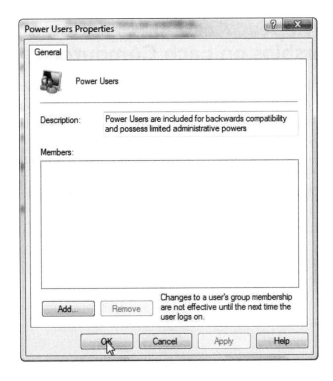

5. Click the Check Names button to verify that the names you entered indeed match actual users.

6. Click the OK button. You should see the username you entered now displayed in the Members section of the Power Users Properties dialog box.

7. Click OK to close the Select Users dialog box, repeat this set of steps to add users as members of other groups, and then close the Computer Management console.

Figure 13-4

The Select Users dialog box allows you to add multiple users to the group at one time.

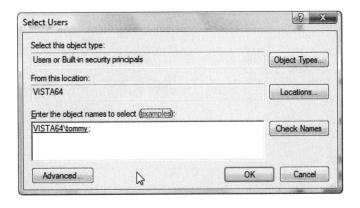

Step 3: Repeat Account Creation and Group Memberships on Each Computer

In a workgroup network with a few computers, you have to reproduce the user account creation process on each and every computer the users will be able to use directly, or have shared file and print access to. When your network grows to six, ten, or more computers, you will begin to appreciate the concept and implementation of centralized servers and managing users and resources from one computer (server) rather than running all over the house to each computer. The same tools you just used to create and manage user accounts are also used to disable or delete users.

Project 14

Control Other
PCs Remotely

What You'll Need

- **Your computers**
- **Your local network**
- **Cost: $0**

Some days you're just too lazy to get up and walk over to use someone else's computer. Windows XP and Vista have built-in remote-control support tools that enable you to pretend you are sitting at another computer—viewing the screen, moving the mouse, typing characters, and running programs. Remote control can save you the cost of extra monitors, keyboards, and cables, and lets you provide assistance to others across your network or the Internet.

Remote control of computers has been around, since, well, forever in computer time—from basic text-only terminal sessions, to the Internet technology bringing us Telnet, to programs like pcAnywhere, to VNC screen-to-screen sessions, to Windows Remote Desktop.

Remote Desktop is part of Windows XP and Vista. It can be used on your local network as well as over your Internet connection to PCs far away. To use Remote Desktop over the Internet, you send a remote control "help" request to another user, by e-mail or Windows Messenger, and the recipient activates the request to log on.

In PC-to-PC use on your local network, Remote Desktop has some limitations—the computer to be controlled must be turned on, booted up, and online, and the controlling computer takes complete control, logging off the user of the computer to be controlled and transferring control to the other computer.

To use the remote computer locally, you first must configure the computer you want to control to accept remote connections from specific users. Then, you simply open a Remote Desktop application on the computer you want to control from, point to the address or name of the computer to be controlled, and log on.

Step 1: Configure Windows to Accept Remote Connections

Configuring Windows XP or Vista to accept remote connections takes four easy progressions through the following dialog boxes:

1. Right-click My Computer and choose Properties.

2. In the left pane of the Control Panel > System window, shown in Figure 14-1, click Remote Settings.

Figure 14-1

Access Remote Settings through the Windows Control Panel.

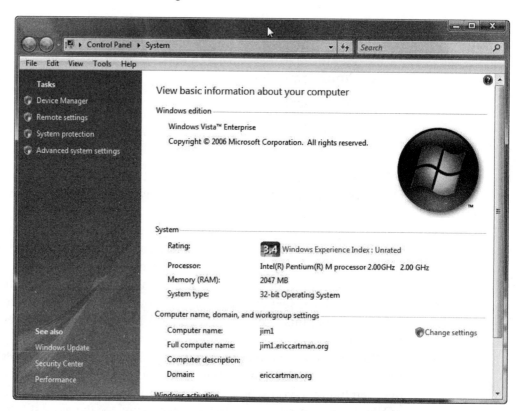

3. On the Remote tab of the System Properties dialog box, shown in Figure 14-2, check the Allow Remote Assistance Connections to This Computer check box and choose the Allow Connections from Computers Running Any Version of Remote Desktop (Less Secure) radio button under Remote Desktop.

4. Click the Select Remote Users button to access the Remote Desktop Users dialog box, shown in Figure 14-3.

5. In the Remote Desktop Users dialog box, click the Add button and type in the name of users who have accounts on this computer—those who you want to be able to take control of this computer remotely.

6. Click the OK button to close these dialog boxes and then close the Control Panel. This computer is now ready for remote-control connections.

Figure 14-2

Enable Remote
Assistance and Remote
Desktop through the
System Properties
dialog box.

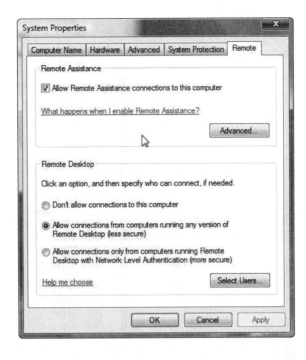

Figure 14-3

Allow specific users
to make remote
connections to this
computer in the
Remote Desktop Users
dialog box.

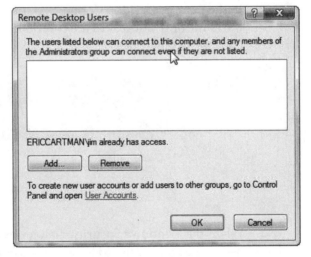

Step 2: Connect to and Control Your Remote Computer

Accessing the desktop of a remote computer is similar to connecting to a shared drive or printer—"address" the computer, log on, and suddenly the desktop interface of another computer appears in a new window on your computer. Follow these steps:

1. Go to the Start menu, expand All Programs, select Accessories (see Figure 14-4), and then click Remote Desktop Connection. (In XP Home the Remote Desktop Connection may appear under the Communications sub-menu.)

Figure 14-4

Access to the Remote
Desktop Connection
application is through
the Accessories
program group.

2. In the Remote Desktop Connection dialog box, shown in Figure 14-5, type in the name or IP address of the computer you want to take control of, and then click the Connect button.

Figure 14-5

Allow specific users
to make remote
connections to this
computer in the
Remote Desktop
Connection dialog box.

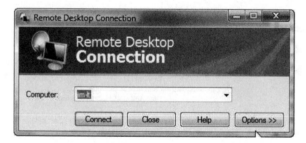

3. When the Windows Security dialog box appears (see Figure 14-6), type in a username and password for a user account that has Remote Desktop permission (from the preceding Step 1 section).

Figure 14-6

Log on to the remote computer through the Windows Security dialog box.

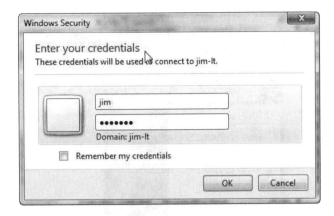

4. Once the connection is established, if another user is logged onto the remote computer, you will see two messages appear in the remote session window. The first message, shown in Figure 14-7, informs you about the other logged-on user and asks you to confirm you want to log them off and take control. The second message simply provides a status message as the other user is logged off.

Figure 14-7

Windows is polite enough to ask you to confirm logging off the other user to let you have control.

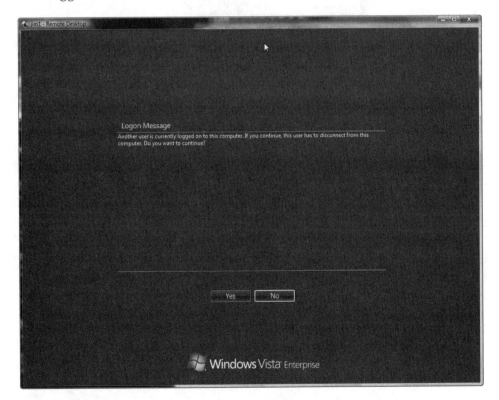

5. When the remote session is established, you see the desktop of the other computer, as shown in Figure 14-8. The response of the keyboard, mouse, and screen will be considerably slower than when you are using your local computer, and if the desktops are set up the same way, it is easy to confuse which computer you are using.

Figure 14-8

Viewing a remote
desktop through a
window on your
local PC

Alternatives to Windows Remote Desktop

One drawback of Windows Remote Desktop is that only the user controlling the remote computer can interact with the screen—that is, both the controlling user and the user at the computer being controlled cannot share the desktop experience. Free programs based on AT&T Cambridge UK's Virtual Network Computing, or VNC, not only offer user-level security equal to or better than Windows security, but also allow both users to interact with the desktop. There are cross-platform versions that let you operate Apple Macintosh, Linux, and Windows systems from any other platform. The following are some programs to try:

- **VNC Free Edition**   www.realvnc.com
- **TightVNC**   www.tightvnc.com
- **UltraVNC**   www.uvnc.com
- **DameWare Mini Remote Control**   www.dameware.com (30-day trial)

Project 15

Stream Your Own Content

What You'll Need

- **Your computer**
- **Your local network and Internet connection**
- **Windows Media Encoder software—free**
- **Audio patch cable with 3.5mm stereo plugs (male connectors) at both ends**
- **Cost: Cable $5**

At some point you may envision having a total e-house (or is that i-house?)—complete interconnection between computers, media devices, telecommunications equipment (phone, fax), and home automation—and providing control of one or more resources from anywhere over the Internet.

Through the projects in this book, we actually get pretty close to a fully-connected and integrated e-house—from single-computer broadband Internet access to webcam and internal host access. All of these projects are basic building blocks that enable you to do more and more with your home network as you learn more and broaden your interests and skills.

On a whim, amidst writing this book, I decided to broadcast what my police/fire scanner hears. I enjoy listening to the somewhat animated dispatchers of the Dallas Fire Department (a routine "Let 'em roll, let 'em roll, Engine 57, Truck 24, Battalion 16…" or an early morning wakeup "Out of bed sleepy-heads something's burning…" brings back memories of my days as a volunteer in Texas) which made me decide to share some of the public-safety communications in and around Silicon Valley, California—not necessarily exciting, but I could fill a void since I could find no one else doing it. I got through with the project and (whack on forehead…) realized, "Hey! Something new for the book!" So here it is, a project that shows you how to stream your own content.

These steps are applicable to scanners (see Figure 15-1), MP3 players, a simple microphone attached to your PC—any external source of audio content or digital content stored on your PC. You may have to get a little creative with some audio cables and connectors, but after a quick software download and installation, in just a few minutes, what you hear can be heard by millions.

Figure 15-1

The audio output of a police scanner or any source of audio can be streamed across your network and to the Internet.

Step 1: Install Windows Media Encoder

The heart of this project is Microsoft's free streaming-media program, Windows Media Encoder (WME). WME is a small but robust piece of software that can serve up on-demand content to Windows Media Player client programs, or it can push audio and video to a streaming media server that sends the program material out to client computers. In this project, we'll configure WME to allow Internet users to connect to an audio stream by typing a URL into Windows Media Player. The first part of this process is to obtain and install Windows Media Encoder, as follows:

1. Go to www.microsoft.com/windows/windowsmedia/forpros/encoder/ default.mspx or just search for Windows Media Encoder to get to the link.

2. Download Windows Media Encoder 9 (32- or 64-bit version, depending on your PC and operating system) and install it.

3. Move on to Step 2 to connect your audio source before you start the program.

Step 2: Set Up Your Content Source

If you are using a police scanner as I am in this example (though any similar audio source—shortwave radio, MP3 player, and so on—will work fine), program the channels or set the stations or playlist as you like. Set the volume control at a comfortable/normal listening level playing through the built-in speaker or headphones. Then, begin the hookup and audio level settings, as follows:

1. At your PC, double-click the speaker icon in your task tray to open the Volume Control panel, shown in Figure 15-2, to access the playback level settings.

Figure 15-2

Windows' normal audio controls are used to set local record and playback audio levels.

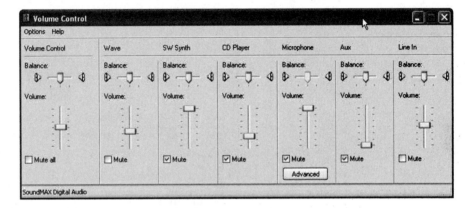

2. Decrease the Volume Control setting to 10–20 percent of maximum, or to a comfortable local PC speaker listening level you use for Windows sounds.

3. Connect one end of the audio patch cable (see Figure 15-3) to the radio/player speaker/headphone jack and connect the other end to your PC's Aux, or Line In, audio input jack.

Figure 15-3

A stereo patch cable quickly connects your audio source to your PC's sound interface.

note *Audio/electronics purists will instantly recognize that the electrical match between your device's speaker output and your PC's audio input are not ideal, and I'm a bit embarrassed to admit to this blatant violation of sound-engineering practices, but you'd be amazed at how well this works (and not blow out your device).*

4. Double-click the speaker icon again to open another mixer control panel.

5. Choose Options | Properties to view the mixer Properties dialog box, shown in Figure 15-4.

Figure 15-4

Windows' audio control panels are selected by the sound card or audio mixer properties.

6. Choose the Recording radio button, check either the Aux or Line In check box, depending on which audio input jack on your PC you connected the patch cable to, and then click the OK button.

7. In the Recording Control panel, shown in Figure 15-5, check only the Select check box for Aux or Line In and then adjust the level to 10–20 percent of maximum.

Figure 15-5

For streaming content, you need to select only one audio source, though you could mix two or more together.

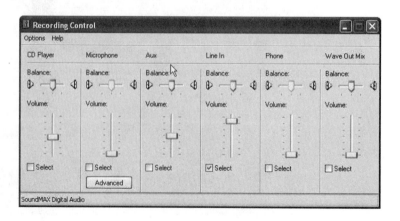

8. In the Volume Control panel, check the Mute check boxes for all of the audio sources except Wave and Aux or Line In. Set the Aux or Line In level control midway.

9. In the Recording Control panel, while listening to a transmission coming from the connected radio/source (scanner), adjust the Aux or Line In level control for a comfortable, undistorted listening level, and then reduce the control level one-half to one full notch. This level should be adequate for streaming.

10. In the Volume Control panel, select the Mute check box for the audio source (unless you want to listen to your scanner live through your PC, but you can do that after we've set up the streaming).

Step 3: Configure Windows Media Encoder

Media Encoder can push or serve many types of multimedia content, from video to stereo audio, live or prerecorded. This project simply encodes and serves up monophonic voice audio—nothing as rich as a CD-quality music library, but a good start on learning about and using Internet audio technologies. There are eight parts to configuring Media Encoder to put your desired content "out there":

1. From the Start menu, expand All Programs and run the Media Encoder to begin initial configuration. Choose Broadcast a Live Event on the New Session screen, shown in Figure 15-6, and then click the OK button.

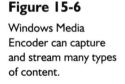

Figure 15-6

Windows Media Encoder can capture and stream many types of content.

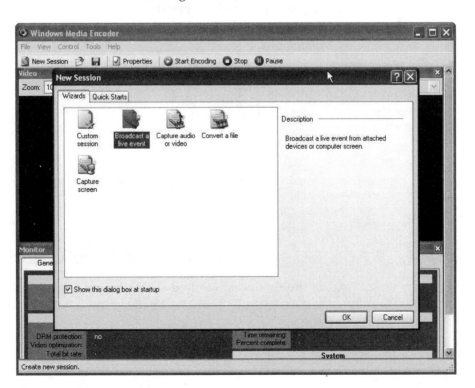

2. In the Device Options dialog box of the New Session Wizard, shown in Figure 15-7, make sure the Audio check box is checked and then click the Configure button.

Figure 15-7

The Device Options dialog box gives you access to select specific details for your audio connection.

3. In the Properties dialog box for your device, an example of which is shown in Figure 15-8, choose from the Pin Line drop-down list the type of connection you plugged the patch cable into—Aux or Line In—and then click the OK button. Click the Next button when you return to the Device Options dialog box.

Figure 15-8

Select the specific audio connection in Mixer Properties.

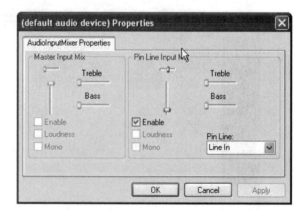

4. In the Broadcast Method dialog box of the wizard, shown in Figure 15-9, select the Pull from the Encoder radio button and then click the Next button.

Figure 15-9

Your stream can be sent to a Windows Media audio server through various online services or played directly from your local PC.

5. In the Broadcast Connection dialog box of the wizard, shown in Figure 15-10, select the specific TCP/IP port to which you want listeners to connect to hear your streamed broadcast. You can pick almost anything, so I chose something catchy and "appropriate" for this example—port 9911. Click the Next button to move to the last configuration dialog box of the wizard.

Figure 15-10

Your media stream should be made available from a unique TCP/IP port.

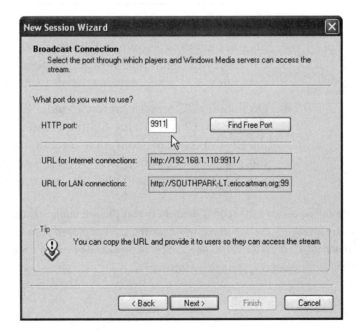

6. In the Encoding Options dialog box of the wizard, shown in Figure 15-11, configure the setting according to the type of audio source you are going to stream. Voice broadcasts such as those from a police scanner or shortwave radio actually sound best when you select Voice Quality Audio; otherwise, you waste bandwidth and stream out more noise than listeners like to hear. The better the quality setting the more bandwidth that is taken up by the digital audio data stream. Click the Finish button and you are one final step away from becoming an Internet broadcaster.

Figure 15-11

The Voice Quality Audio setting is perfect for normal speech radio broadcasts.

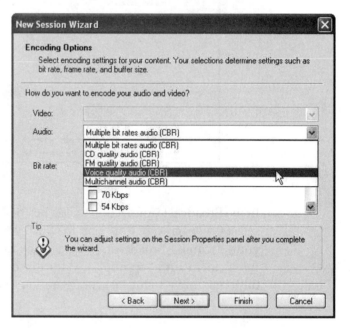

7. Start your broadcast by clicking Start Encoding in the main Windows Media Encoder window, shown in Figure 15-12. A new Audio level meter appears, enabling you to monitor how many people are listening to your stream, the data rates, and other factors relative to streaming audio.

8. On another PC, open Windows Media Player, and press the CTRL+M keys to display the Classic Menu. On the menu select File, then Open URL, type in the IP address of the computer sending the stream, followed by a colon and the port number configured in Step 5, such as 192.168.1.1:9911. From this point, Media Player will pick up your streamed programming.

note There will be about a 10-second delay between the live audio you can monitor at the scanner and streaming PC, and when you will hear the stream played back at another PC. This is completely normal, because WME needs a bit of processing time to encode the source material and buffer it before sending it on to a listening PC.

Figure 15-12

Monitor your stream's
audio levels, network
performance, and
listeners in the Media
Encoder Audio and
Monitor displays.

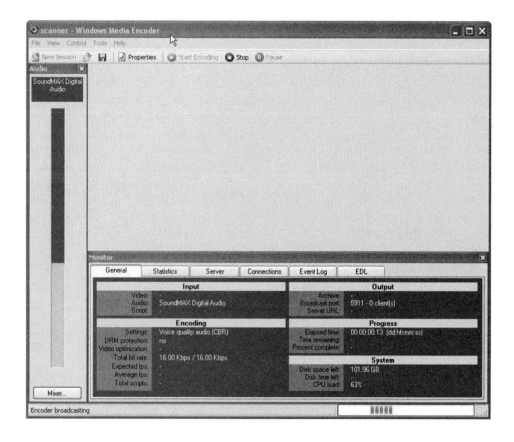

Step 4: Stream Automatically at Startup

You can save your streaming settings and have them start automatically through a
Windows shortcut or batch/command file:

1. In the Windows Media Encoder main window select File | Save As.

2. Pick a folder location and filename for the stream configuration file—for this
   project I used SCANNER.WME. Save the file.

3. Create a batch/command file, or a new Windows shortcut with the following
   command-line content to use as the instant start of the encoder and stream:

   ```
   "C:\Program Files\Windows Media Components\Encoder\wmenc.exe"
   "c:\scanner.wme" /start
   ```

note   *Do not stream/rebroadcast content that you do not have license or rights to distribute. As you
may know from following various cases of media piracy against Napster, YouTube, and MySpace,
copyright laws are very strict. Police, fire, ambulance, amateur radio, marine radio, and so forth
are non-copyright material in the public domain. Such content may not be used for profit.*

Project 16

Supreme Cellular Data Connection Sharing

What You'll Need

- Your computer
- A cellular data card
- Cellular data service
- Kyocera KR1 Mobile Router
- Cost: Cell data card, free to $130; Cellular data services $11–70/month; KR1 Router $179

Project 7 described how to establish a cellular data connection. If you decided that was the best option for you, either because that is the only way you can get to the Internet at home or because you prefer a cellular data connection when you're on the road, this project will help you kick it up another notch and share that connection with other computers. By employing a mobile router, such as the Kyocera KR1 Mobile Router or a similar unit from Linksys or Top Global, you can expand your cellular connection to other computers easily by Ethernet cable or Wi-Fi—at home or when mobile. In this project, we're creating an expanded cellular data Internet configuration that will look like Figure 16-1.

A mobile router can use either the tethered cellular modem Internet access from your cell phone or a PC Card plugged into the router. From there the router acts like any other broadband router/firewall with Wi-Fi access. The details we've covered in previous projects for routers and firewalls give us a good foundation to build on—substituting cellular for cable or DSL. An advantage of cellular connectivity over Wi-Fi

Figure 16-1

Using a cellular data router can extend your Internet connection to cover your entire home, or provide connectivity on the road.

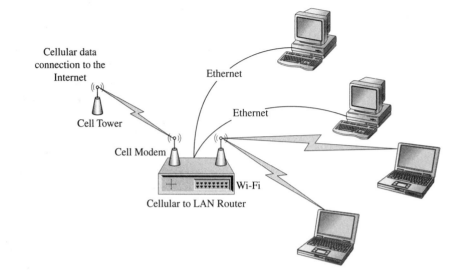

is that it is designed to work on-the-move, so if someone you're traveling with needs to be connected, cellular is the best option available.

Step 1: Set Up the Mobile Router

This project starts with a Kyocera KR1 Mobile Router, shown in Figure 16-2. This router accepts an EVDO (Verizon or Sprint) cellular modem PC Card or USB-tethered data-capable cell phone (of which only a few models are supported so far) as the Wireless Wide Area Network (WWAN) Internet connection. The WWAN connection is equivalent to a cable or DSL connection, though slower.

Figure 16-2

The Kyocera KR1 resembles most other router/firewalls, providing the Internet connection through a cellular data modem.

The KR1 translates the WWAN connection into four Ethernet ports and a Wi-Fi access point (see Figure 16-3). Internally, the router provides Dynamic Host Configuration Protocol (DHCP) routing and firewall functions just like a typical home network router/firewall. With this functionality, the router will support multiple computers on your home network, and let you take your Internet connection on the road.

Figure 16-3

The KR1 Mobile Router has four Ethernet ports and a Wi-Fi access point.

Setting up the KR1 is also similar to configuring a home router, substituting cellular connection parameters for cable or DSL parameters; you access and follow the built-in setup wizard:

1. With the power cord disconnected from the KR1, insert your cellular data modem PC Card into the appropriate slot on the KR1.

2. Connect an Ethernet cable between a computer and one of the RJ-45 Ethernet ports at the back of the KR1.

3. Connect the power cable to the KR1 and let it start up—this takes 1–2 minutes.

4. Access the settings for your Ethernet network connection through the Start Menu, select Control Panel, and double-click Network. Right-click your Ethernet connection and select Properties. Double-click Internet Protocol (TCP/IP) to access the TCP/IP Properties, select the Obtain An IP Address Automatically and Obtain DNS Server Address Automatically radio buttons, and click OK to close the dialog boxes.

5. Access the KR1's setup home page, shown in Figure 16-4, by opening Internet Explorer and typing the IP address **192.168.0.1** into IE's Address bar.

Figure 16-4

The KR1 mobile router provides internal web-based configuration.

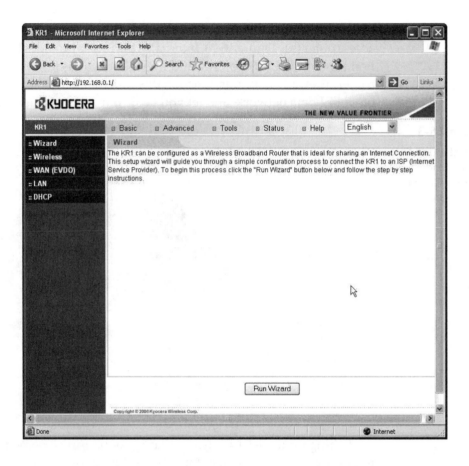

6. On the KR1 router's home page, click the Run Wizard button.

7. The first page of the setup wizard (see Figure 16-5) prompts you to change the access password for the device from the default, admin. This is a good time to do this and is highly recommended. Click the Next button to continue.

Figure 16-5

Changing the password on your router is a high priority as part of setup.

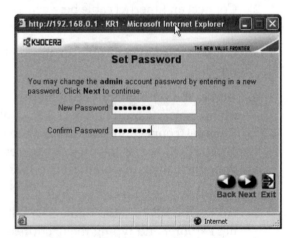

8. On the next wizard page (see Figure 16-6), select your time zone. This is more a convenience for accurate log files and not a requirement. Set the time zone and then click the Next button.

Figure 16-6

A correct time zone setting helps you reconcile connection times and log files.

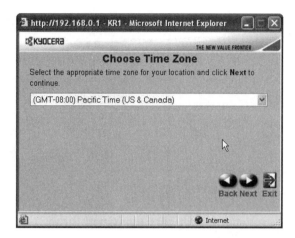

9. The wizard prompts you for WAN settings (see Figure 16-7), in this case, the settings necessary for your router and cell modem to make a data connection. For most EVDO connections, all fields are left blank except the Dial-Number field, which is usually #777. You don't need to change the default settings, so click the Next button.

Figure 16-7

Most EVDO connections do not require any configuration changes from the default settings.

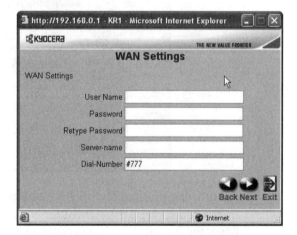

10. The next step of the KR1 setup wizard (see Figure 16-8) is to establish the Wi-Fi settings for the wireless portion of the router. Type in the name (SSID) you'd like to give the router's wireless network, select a channel to use (Channel 6 is the typical default selection for most Wi-Fi gear—choosing Channel 1 or 11 will usually keep you away from others' interference), and then click the Next button.

Figure 16-8

The KRI's wireless network settings are typical of any Wi-Fi network.

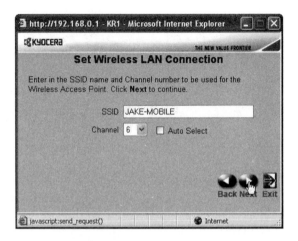

11. On the Setup Encryption page (see Figure 16-9), provide the security settings you would like for your wireless network—WPA and a complicated passphrase are recommended. Click the Next button.

Figure 16-9

Adequate security settings are critical for Wi-Fi networks.

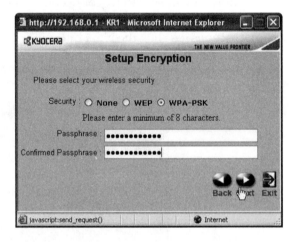

12. At the completion of the KR1 setup wizard (see Figure 16-10), you are instructed to restart your router to save and apply the settings you chose. Click the Restart button to continue.

Figure 16-10

Restarting your KRI router saves and applies the changes you made.

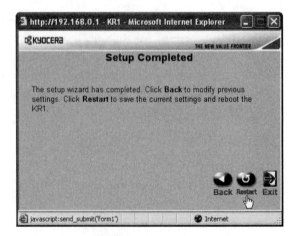

Step 2: Restart and Verify the Mobile Router Configuration

After restart, your KR1 router is ready to connect your computers to the cellular data network. To conserve airtime and protect the use of your cellular service, you must make the cellular connection manually.

Access the KR1 home page, log in, and then click the Status button in the top navigation area. On the Device Info page, shown in Figure 16-11, you will see the status of the LAN, WAN, and wireless (Wi-Fi) connections. The WAN section of the page is where you control the cellular data connection (with the Connect and Disconnect buttons) and see the status, including the cellular IP address and gateway, and DNS information which will be passed on to the client computers—there is no need to configure this information. Since mobile routers are specific to Verizon or Sprint EV-DO systems, they come preconfigured to connect to their respective networks depending on the cellular data card or cellular phone you provide.

As Internet connections go, cellular is certainly not in the same league for speed as cable or DSL services, but it can be consistent, reliable, and available in less-populated areas where cable and DSL providers have yet to make their services available. Sharing your cellular connection with a mobile router—and, yes, it will work in the car with a DC power adapter—means you don't have to share your computer while it keeps you and others in touch on the road.

Figure 16-11

The KR1's cellular connection is controlled from within the Device Info web page.

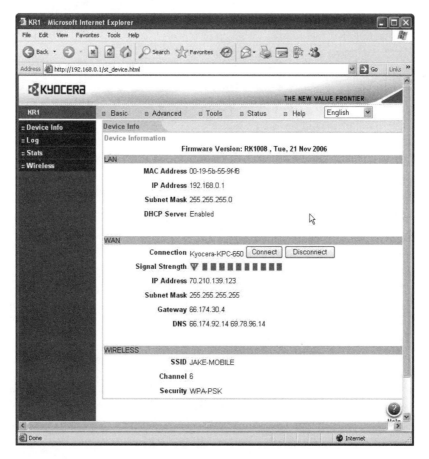

Project 17

PC Weather Station

What You'll Need

- Your computer
- Internet connection
- A PC-enabled weather station
- Common hand tools, a drill motor and drill bits
- Citizen Weather Observer Program or Weather Underground registration
- Cost: Weather station $120–400; CWOP and Weather Underground registrations, free

A fun and informative project is to gather your local weather information and share it with others, because the weather around the globe is as big as or bigger than the Internet. Anywhere you live or travel, the weather is an important factor of life, work, recreation, chores around the house, and vacation trips to far-away places.

Creating a PC-based weather station that reports to the Internet requires, of course, a weather station that can connect to a PC, and an hour or two of time to install, connect, and configure. I found all manner of moderate-to-expensive weather station systems from Davis Instruments (www.davisnet.com), La Crosse Technology (www.lacrossetechnology.com), Oregon Scientific (www.oregonscientific.com), and others, but for a quick, economical start I was glad to find the 1-Wire Weather Instrument Kit V3.0 from AAG Electronica (www.aagelectronica.com) for $120. Some assembly is required with the AAG weather station, so you may find the prebuilt units more to your liking. We'll assume our weather station is assembled and ready to install.

Step 1: Verify Your Weather Station to PC Connections

Since the measuring unit of a weather station is best placed at some height outside your home, you should make sure the software and cable connections function correctly before stringing a cable and climbing on the roof to secure the measuring unit.

Download (if necessary) and then install the driver and application software for your weather station. The AAG weather instrument comes with a short cable and an interface adapter to connect to your computer—make the connection between the measuring unit and your PC, and then test the software (see Figure 17-1).

Figure 17-1

Weather station
software configuration
for the AAG instrument

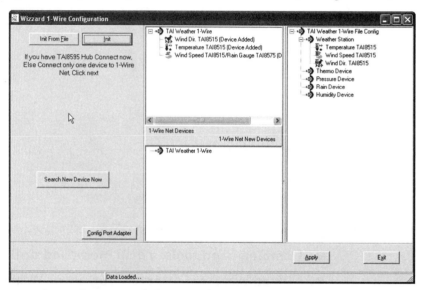

Once you have determined that you have temperature, wind speed, and direction measurements displaying on your PC, as shown in Figure 17-2, you're ready to prepare cabling and install the measuring unit where it will work best.

Figure 17-2

The AAG weather
station software
showing data from the
measuring unit

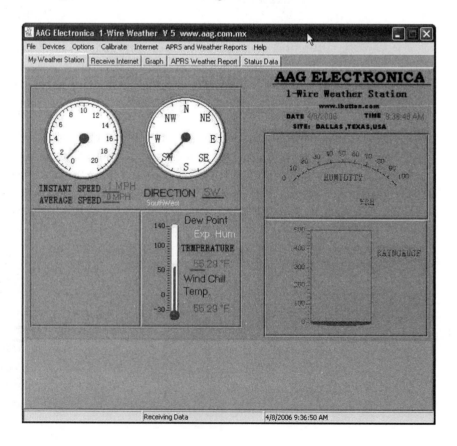

Step 2: Prepare the Cable Connections

Because the AAG weather station uses standard phone cable and connectors between the sensor and PC, the signals are such that they can easily get lost in the cable, especially a long (25 feet or more) cable. If you are going to place the sensor unit on a 16-foot-tall pole, or your roof, the cable length needed will probably be at least 25 feet.

To avoid the loss of signal (which appears as "no device detected" in the test programs) you need to modify the standard phone cable so it can pass the data signals over a long distance. The trick is that normal phone cables use two wires parallel to each other, which creates a lot of capacitance and diminishes that sensor data signal. To avoid this capacitance, we need to use nonadjacent wires.

The necessary cable modification consists of removing the original connectors from a premade cable and reattaching connectors using different internal wires, or making your own special cable. It's very easy to do this modification using a $6–10 RJ-11 connector crimping tool and a pair of RJ-11 modular plugs.

First, cut off and discard both of the original connectors on the phone cable you have, as shown in the example in Figure 17-3. Next, using the insulation stripping blade on the crimping tool, remove the insulation from both ends of the cable. This will expose a black, red, green, and yellow insulated wire.

Figure 17-3

Modifying a standard phone cable to work with the weather station

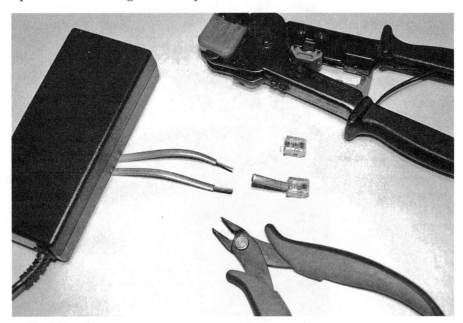

In a normal phone cable the middle contacts use the red and green wires, carrying the signal, and the same is true for the 1-Wire Weather Sensor—the two middle contacts are the signal connections. Because the 1-Wire system does not communicate well using red and green wires which are next to each other we need to change the connector wiring. To make this change we need to pick two nonadjacent wires—either black and green or red and yellow—and have them connect to the middle two contacts in the connectors at both ends of the cable (we do not need the other two wires to connect to anything).

I picked the red and yellow wires, cut off the green and black wires, inserted the red and yellow wires into the same middle two contact pins in the connectors at each end, and then crimped firmly. When you do this, be sure the contacts used for the red and yellow wires are the same at each end. Reversing them will cause the connection to fail.

When the connectors are attached, you may repeat Step 1 to verify that the connections are correct and the cable works. When successful, you are ready to prepare the mounting and install the sensor unit.

Step 3: Install Your Weather Station

The measurement units of most weather stations can typically be placed 100–300 feet away from the computer (or display unit). For best results, the measuring unit or head of the weather station should be mounted in a clear location at least 16 feet above ground level, and 4 feet or more above a roof or deck, with clear exposure in all directions—to get accurate wind speed and direction measurements. String the cable and secure it with U-shaped wire staples or plastic clips like those used for a TV antenna cable.

The 1-Wire Weather Station should be mounted at least 16 feet above the ground or rooftop and away from surrounding objects to provide the most accurate wind and temperature readings. Although generally accurate mounted at eye level, or as far as you dare reach above a rooftop mounting (as I have done, as shown in Figure 17-4), summertime temperature readings may be a couple of degrees higher than ambient—but above 90 degrees, what difference is a couple degrees of "scorching" really going to mean?

Figure 17-4

The AAG weather station mounted 5 feet up a roof-mounted TV mast, ready for use

Warnings/Cautions

Although building this project is generally a safe endeavor, you do need to mount the weather sensor at least 16 feet above ground—or a few feet above roof level. Proper care and safety precautions should be taken when using power tools, climbing ladders, and working on the roof or on any elevation.

When working with ladders, poles, wires, and mounting the sensor unit, stay as far away (generally 3 feet or more) from power and other utility lines as possible to avoid electrical shock and damage.

During electrical storms, to avoid damage to your PC, disconnect the cable from the sensor unit at the back of your PC.

Step 4: Configure Weather Software to Send Data to the Internet

Once you have registered and have logon information for either CWOP (www.wxqa .com) or Weather Underground (www.wunderground.com), or both, you need to configure the weather station software to send your weather data:

1. On the menu bar of the AAG program, select APRS and Weather Reports menu item.

2. On the WSI Location tab, enter the latitude and longitude of your location, and then click the CWOP tab.

3. On the CWOP tab (see Figure 17-5), enter your station ID (your amateur radio call sign or your CWOP ID number). You can usually leave the Host and Port settings at their default settings. Click the Wunderground.com tab.

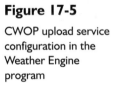

Figure 17-5

CWOP upload service configuration in the Weather Engine program

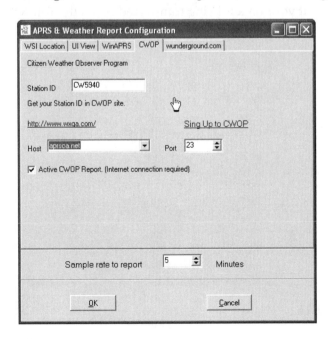

4. On the Wunderground.com tab (see Figure 17-6), enter your station user ID and password in the corresponding fields, and then check the Active Wundergroup Report check box. Click the OK button to close the dialog box, and then wait 5–10 minutes for your station to report over the Internet.

Figure 17-6

Weather Underground upload service configuration in the Weather Engine program

When the weather station software is properly configured, it will upload your data to either or both weather services for you and others to observe the records of your local weather. Log onto each weather reporting site, www.wunderground .com (see Figure 17-7) and www.findu.com using your weather report accounts (click Weather Reports to reach a page similar to the example shown in Figure 17-8), to see if your data is being transferred correctly. You will notice that both sites present your location in latitude and longitude and can present maps of your location.

Figure 17-7

The Weather Underground service provides a daily histogram of your weather conditions and can show you data from other weather stations near you.

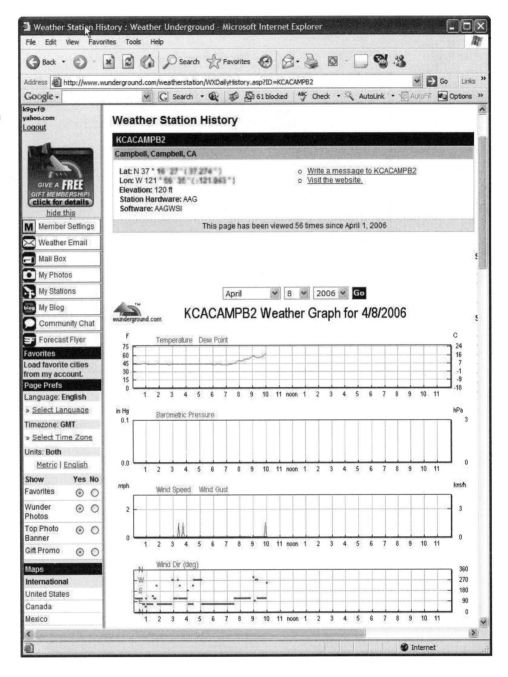

Figure 17-8

Findu.com provides a basic quick look at your weather station data, including a log of raw data transmitted over time.

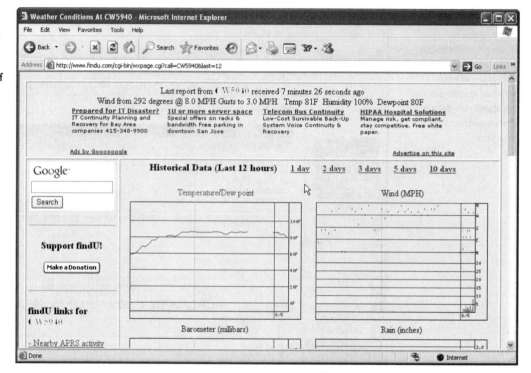

Project 18

SlingLink TURBO

What You'll Need

- Your computers
- Your home network
- SlingLink TURBO unit
- Cost: SlingLink TURBO 1 Port $100 (4 Port $130)

This book has shown a lot of ways to mix and enjoy multimedia on your home network—from wired to wireless—and now there's a hybrid of wired and wireless made just for those locations to which you cannot run wires and in which you don't want to mix together wireless devices. The SlingLink TURBO (actually a pair of units; see Figure 18-1) from Sling Media (www.slingmedia.com) uses your existing AC power lines to convey Ethernet from one location to another—perfect for getting to the back of your media equipment.

Figure 18-1

The SlingLink TURBO pair provides an easy extension or connection to your home network where wires and wireless are not suitable.

Only two connections are required—standard AC power, and Ethernet. The SlingLink TURBO acts as an invisible bridge, from anywhere on your home network (but typically your router) to a Slingbox AV (see Project 9), another computer, a wireless access point—anything that needs a network connection apart from your original wired network. The schematic of this network connection is represented in Figure 18-2.

Figure 18-2

The SlingLink TURBO pair forms a network bridge through normal power lines.

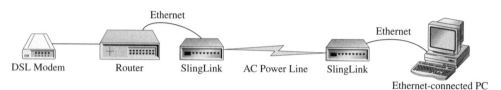

No configuration is necessary—no IP, gateway, or DNS addresses to set. It's literally plug and play. Due to technical limitations of using power lines for data transmission, your data throughput (bandwidth) is not as fast as most direct-wired Ethernet connections, so the SlingLink TURBO is not for those requiring the full bandwidth of their 100 Mbps or Gigabit Ethernet connection for local network performance. However, it passes the typical DSL download speeds just fine.

Some typical uses for the SlinkLink TURBO are as follows:

● Providing short links from your cabled network to media centers to connect your DVR (such as TiVo) to the Internet

● Broadcasting your Sling Media output or your TiVo files to another PC

● Extending your LAN to PCs you cannot reach with wires or wireless connections due to obstructions or physical limitations

● Extending your wired LAN to corporate PCs when you are not allowed to reconfigure network settings or add wireless support

● Providing a quick lash-up for gaming consoles

Project 19

Firewall Configurations for IM and Voice over IP

What You'll Need

- Your computer
- Your router/firewall and firewall software
- Broadband Internet connection
- Your choice of IM or VoIP software
- Cost: $0

Instant messaging isn't just about text messaging…and hasn't been for a long, long time. Today we can use AOL Instant Messenger (AIM), Windows Live Messenger, and Yahoo Messenger to share live webcam images, computer-to-computer voice conversations, and computer-to-phone calls.

Adding to the diversity of our computers are Voice over IP (VoIP) applications such as Packet8, Skype, and ViaTalk for computer-to-computer, computer-to-phone, and phone-to-computer calls. Computer-based phone service is a relatively new but increasingly popular means to connect to the huge domestic and international phone systems, or simply chat between PCs.

Although all of these applications provide the same basic features—voice over the Internet—they require unique configurations of both software and hardware firewalls. This project gives you examples of firewall configurations that you need to set for common VoIP, IM, remote control and video applications so that they will work properly through your computer, network, and broadband connection.

Port and Protocol Definitions

As you've explored networking, you've undoubtedly run across myriad new terms—from IP address to protocol, TCP to UDP, and so on. These terms all refer to critical elements of connecting a computer to another computer, connecting a computer to a server (web, FTP, or e-mail), or connecting computers to a server through firewalls and routers. The following simple definitions are worthy of further investigation if you encounter connectivity issues or want to become a networking expert:

- **Media Access Control (MAC) address** A unique number given to each and every piece of networking hardware, keyed by equipment manufacturer. MAC addresses are typically not relevant to basic networking functions.

- **Internet Protocol (IP) address** A set of four numbers (192.168.1.2) separated by decimal points, that represents a specific host (computer, server, or appliance) on a network. There are two classes of IP addresses:

  - **Public or routable addresses** Can be used anywhere across the Internet or internal networks. Public IP addresses are provided by network administrators and top-level Internet domain managers.

  - **Private or unroutable addresses** Cannot be used across the Internet. They are in specific numerical groups, typically 10.x.y.z, 169.x.y.z, 172.x.y.z, and 192.168.y.z. Private IP addresses are commonly used by consumer routers and firewalls. Routers and firewalls translate internal/private network addresses to public addresses (the ones used by your ISP) so that your home network data traffic can come and go across the Internet.

- **Hostname** A user-friendly name that translates to a specific IP address, so we can focus on names of computers instead of complex arrays of numbers. Hostnames are translated into IP addresses in the background by Domain Name System (DNS) or Windows Internet Naming Service (WINS).

- **Domain Name System Server (DNS Server)** A local or Internet-based server that looks up host and domain names and returns an IP address for an application to connect to.

- **Gateway** A function or device that determines the best path for a requested piece of information to follow. Your router keeps all local network traffic "behind the gateway," while requests for data across the Internet are sent and come back through the gateway.

- **Protocol** Essentially, a unique language of a specific type of data communications, or how the communications is structured. Two protocols are most prevalent across the Internet, TCP and UDP. Most data types—HTTP, FTP, HTTPS, POP, and SMTP—use TCP because TCP ensures data delivery and integrity from end to end. Data types such as streaming audio and video typically use UDP because they need to transfer quickly yet can suffer some loss without losing the essence of the content.

- **User Datagram Protocol (UDP)**   A quick method of sending short packets of information to another computer. There is no guarantee that the message will arrive, or arrive intact.

- **Transmission Control Protocol (TCP)**   The most common data transmission method used on networks and the Internet—data messages of various lengths are guaranteed (within reason) to reach their intended destination. Multiple applications use TCP protocol, but those applications are subdivided into different sessions or port numbers for each application or computer/host.

- **Ports (TCP, UDP)**   May be thought of as tunnels or channels within a data connection between computers. There are 65,536 possible ports for each of the TCP and UDP protocols. When a single computer is communicating with multiple servers or computers, only one channel may be used per application (web browser, e-mail client, IM, streaming audio or video, and so forth). While all that data is going to and from one computer (and the one IP address for that computer), ports and protocols allow multiple data streams to flow.

- **Services**   HTTP (web browsing), FTP (file transfers), POP3 (e-mail retrieval), and SMTP (e-mail transmission) are all forms of data services. These and many IM services use the TCP protocol to ensure complete data requests and responses between computer and server. Voice chat and webcam services typically use UDP.

Even if you don't embrace these terms and their significance, at least you'll be familiar with their application and place in the configuration of your network and the applications that communicate through it.

Firewall Settings for Popular IM and VoIP Services

In most cases of home networking, you won't have to touch a thing—unless your software firewall prompts you to authorize a specific application to connect, or your application requires an inbound connection, which are usually blocked by both hardware and software firewalls. It is when things do not work that you have to find out what your applications and services need and dig into your firewalls and open things up.

Firewall Settings for Yahoo Messenger, Webcam, and Voice Services

Yahoo Messenger text messaging automatically uses TCP port 5050, but can function on port 80 or 443. The webcam sharing feature uses TCP port 5100. Yahoo Voice Chat uses TCP or UDP ports 5000–5010. Yahoo Voice service uses TCP ports 20, 23, 25, 80, 119, 5050, 8001, and 8002.

The Yahoo Messenger client does not allow you to change any of these settings to work around proxy server or firewall restrictions, but it does allow you to change its configuration to suit a proxy server (common in most corporate networks) or

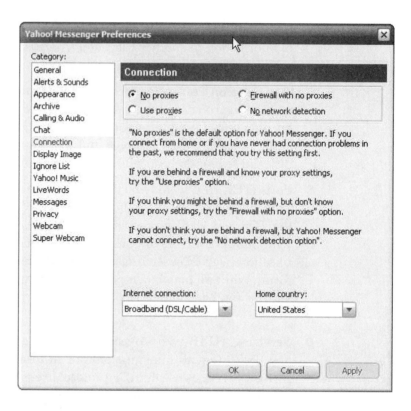

Figure 19-1

Yahoo Messenger will accommodate proxy server and firewall connections to the Internet.

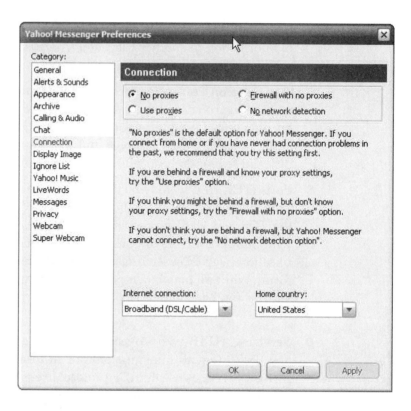

firewall—see Figure 19-1. To access the Yahoo Messenger Preferences window, choose Messenger | Preferences | Connection.

If Yahoo Messenger does not work, your options are to configure proxy server settings if you know there is a proxy server between your computers and the Internet (mostly at work, rarely at home) or configure your firewall to allow access for the ports needed.

Firewall Settings for AIM

AIM's settings are about as simple as you can get: it offers no network configuration settings—it simply figures out the best port to use on its own. It works well through proxy servers and firewalls, preferring to use TCP ports 5190–5193, but basic text messaging is also known to use ports 23 and 80 in a pinch.

Firewall Settings for Microsoft Live Messenger

Leave it to Microsoft to provide one of the most complex (in terms of its communications) messaging applications available. Live Messenger can be configured to work through a firewall, as shown in Figure 19-2, but that is all you have available to configure to get it to work. If you find that you need to configure your firewall to allow Live Messenger to make outbound connections or accept inbound connections, with all of the ports used by all of the available features, you'll be pretty busy.

Figure 19-2

Live Messenger offers
a simple connection
settings interface as well
as testing features to
help you get connected.

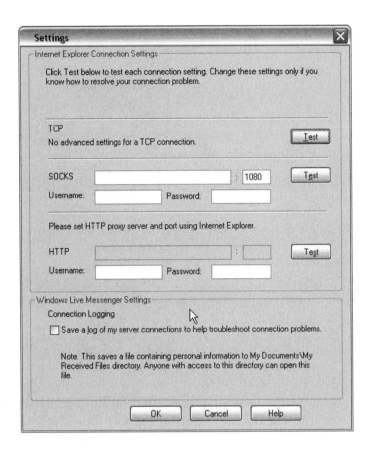

The text chat feature prefers TCP port 1863, but is known to function on ports 80, 443, 1493, 1542, and 1963. The file transfer/sharing feature uses TCP ports 1544 and 6891. Remote Assistance uses TCP port 3389. Audio chat uses TCP ports 1556, 11771, and 13803 as well as UDP ports 5004–65535. The Remote Desktop and whiteboard features use TCP ports 389, 522, 1503, 1720, and 1731. Launching games uses TCP port 80. The video conference feature uses TCP ports 9000–9999 and UDP ports 5004–65535. Signing in uses TCP port 443.

Firewall Settings for Skype

Skype (see Figure 19-3) is one of the most popular stand-alone VoIP programs and provides a good example of the typical network configuration for making Internet and normal phone calls from your computer.

The only configuration option you have for Skype is the port to be used for inbound call requests from others, as shown in Figure 19-4. You can access the network configuration in Skype by choosing Tools | Options | Connection. If the Skype service cannot connect on this randomly assigned (at time of installation) port, it will try ports 80 and 443. If you open the indicated port number, or another of your choice, on your firewall, Skype should be able to use it for inbound calls.

Figure 19-3

Skype is a popular VoIP application with an intuitive user interface.

Skype is flexible and effective but recommends opening both outgoing and incoming UDP ports above 1024 (1024–65536). Skype will work through a proxy server, and uses the proxy settings, if any, that have been configured in Internet Explorer.

Figure 19-4

Skype's limited network configuration options make it simple to get your call through.

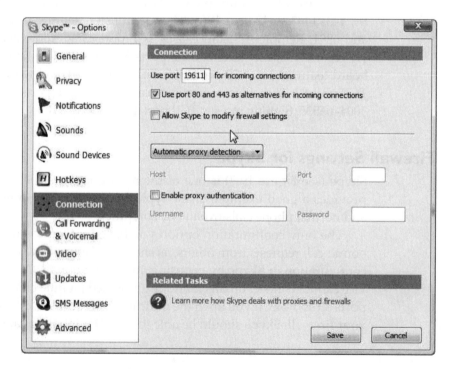

Firewall Settings for Packet8

Packet8 uses several outbound UDP ports, 5060–65534. Few if any hardware firewalls restrict outbound connections—anticipating that you want to make these connections—but software firewalls may be suspicious of outbound connections they do not "know" about. If you use a software firewall, and Packet8 fails to work, you need to configure your firewall to allow the outbound UDP connections for the port range 5060–65534. Packet8 does not request inbound connections, so no configuration to accommodate these is necessary.

Firewall Settings for ViaTalk

ViaTalk seems to use a broad mix of ports for its VoIP services, all of which require inbound connections or port-forwarding to the computer running the ViaTalk client. In addition to normal web TCP port 80 outbound connections, ViaTalk requires inbound UDP configuration for ports 69, 5060, 5061, and 10000 to 20000.

Resolving Firewall Issues

As you encounter potential firewall issues with any application, you can rest assured that someone else has encountered the same or similar problem with the application or the firewall, and can explain a fix or, worse, indicate unquestionably that your particular mix of firewall and application will not work together. You should investigate applications or products you are considering purchasing by searching out reviews and support issues at sites such as CNET.com, and of course searching Ask. com, Google.com, and other reference sites.

One tricky item to note for any application that requires inbound port access through your firewall: the inbound port access in most hardware firewalls allows that inbound connection to one and only one computer/IP address. Therefore, with typical consumer firewalls, allowing more than one computer access to and use of such applications is very difficult, if not impossible. You are better off choosing applications that can function with and rely on only outbound connections.

Part III

Challenging

Project 20

Bridging a Gap, Wirelessly

What You'll Need

- A wireless bridge—aka gaming adapter
- Your computers with web browsers
- Cost: $30–60 for wireless gaming adapter

For some of us, it is relatively easy to expand our network with wires. We "simply" crawl under the house or in the attic, run some new wires, drill a few holes, install a couple of RJ-45 jacks, plug in, and go. For many people—apartment dwellers, house renters, locations with impenetrable ceilings and walls—expanding a network connection that requires wires at both ends is at least a challenge, if not impossible.

While a laptop may network wirelessly over relatively short distances, walls, pipes, electrical cabling, and distance can inhibit connectivity. Taking an example from popular gaming consoles which have only a wired Ethernet port, some computing devices such as network-capable printers, simply do not connect and network without an Ethernet cable. A solution is to use Wi-Fi hardware to extend your network as far as you can without wires, and then return to wires using the wireless bridge where necessary.

Typically, you would specifically look for a pair of wireless bridges—devices designed to "bridge the gap"—but these devices are usually not in stock at even the most geeky of electronics stores. If you already have Wi-Fi capability at one end of your network, you can bridge the gap by adding just one easy-to-find device—a wireless gaming adapter. As much by coincidence as luck in picking a low-cost solution to try, I discovered that the D-Link DGL-3420 Wireless 108AG Gaming Adapter, shown in Figure 20-1, will stretch my LAN.

Figure 20-1

The D-Link Gaming Adapter is a small but powerful addition to your home network.

By adding a wireless gaming adapter as a network bridge, you can extend your network well beyond just a length of wire or the range of wireless to include other wired PCs distant from the main connection point, similar to the network schematic shown in Figure 20-2. This too can be expanded on with additional bridges, as described in Project 21.

Figure 20-2

Schematic of your expanded, bridged, network

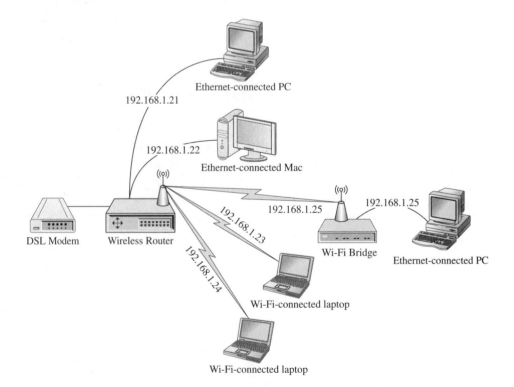

Step 1: Configure a Wireless Gaming Adapter to Become Your Network Bridge

The configuration of your gaming adapter involves first, a connection to your network, or at least directs to a computer to access the web pages built into the adapter. These pages allow you to configure the wireless network system and security settings so the adapter acts just like any other Wi-Fi client device on your network. Once the adapter can connect to your Wi-Fi network, it passes that connection information to whichever device is attached to the Ethernet connection on the adapter. The configuration of the gaming adapter goes as follows:

1. Connect your PC to the gaming adapter by using a single Ethernet cable.

2. Change your computer's Ethernet address to 192.168.0.2. To do so, click Start, right-click My Network Places, and choose Properties. In the Network Connections window, right-click your wired network adapter and choose Properties. In the Properties dialog box, double-click the Internet Protocol (TCP/IP) entry in the list, and then type the addressing information as shown in Figure 20-3. Click OK twice to close the dialog boxes and let the setting take effect.

Figure 20-3

Changing the IP address of a computer to allow access to a new network device

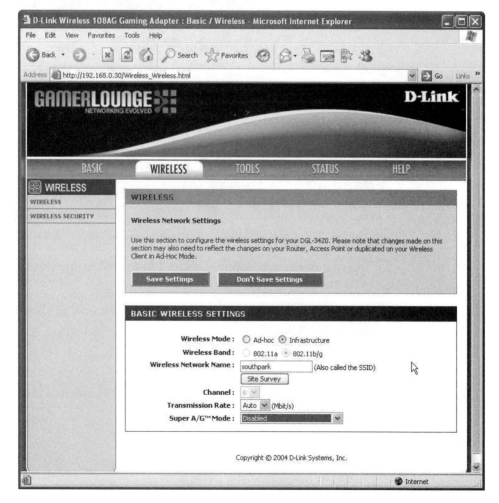

3. In your web browser, type the IP address of the gaming adapter (192.168.0.30 for the D-Link Gaming Adapter) to access the main configuration web page for the adapter. Figure 20-4 shows the page for the D-Link Gaming Adapter.

Figure 20-4

The main configuration web page of the D-Link Gaming Adapter

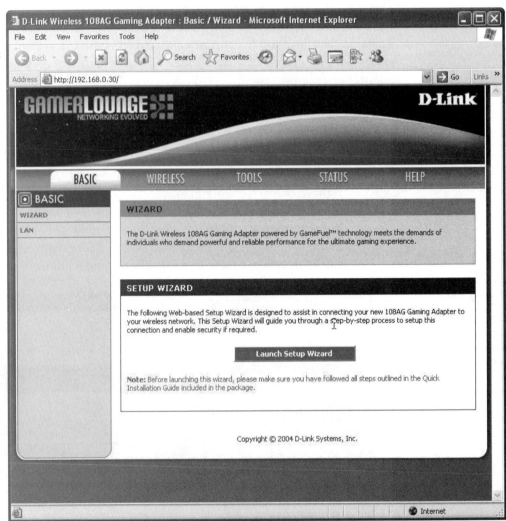

4. Click Launch Setup Wizard. On the Welcome page, shown in Figure 20-5, click Next.

5. In Step 1 of the wizard, shown in Figure 20-6, choose Infrastructure mode and click Next. (In Infrastructure mode Step 2 is skipped and you go to Step 3 next.)

Figure 20-5

The beginning of the D-Link Gaming Adapter configuration sequence

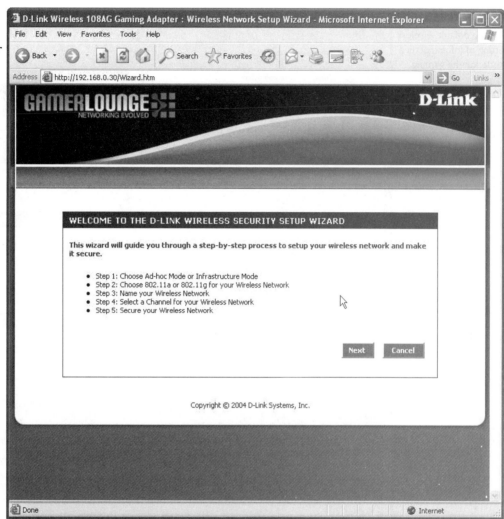

6. Step 3 of the wizard, shown in Figure 20-7, requests your SSID. Enter the SSID of your wireless network and click Next. This will take you to Step 5 of the wizard.

7. At Step 5 leave the default WEP: Disabled radio button selection as-is and click Next. This will take you to the Setup Complete page.

8. When the Setup Complete page appears, click Save. This will re-start the gaming adapter. Wait a minute or two then refresh your browser.

Figure 20-6

Use the default wireless setting of Infrastructure mode for the gaming adapter.

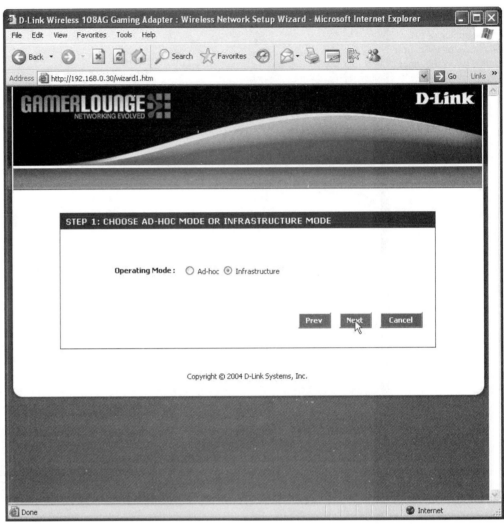

9. After the gaming adapter has restarted, select the Wireless tab at the top of the page, then click Wireless Security on the left menu.

10. On the Wireless Security page, shown in Figure 20-8, choose WPA-Personal for the security mode, choose TKIP from the Cipher Type drop-down list, and type the security key used by your wireless access point in the Pre-Shared Key field.

Click Save Settings in Figure 20-8. You're just a couple of steps away from a complete wireless bridge.

Figure 20-7

The wireless network name on your gaming adapter must match your existing Wi-Fi network name.

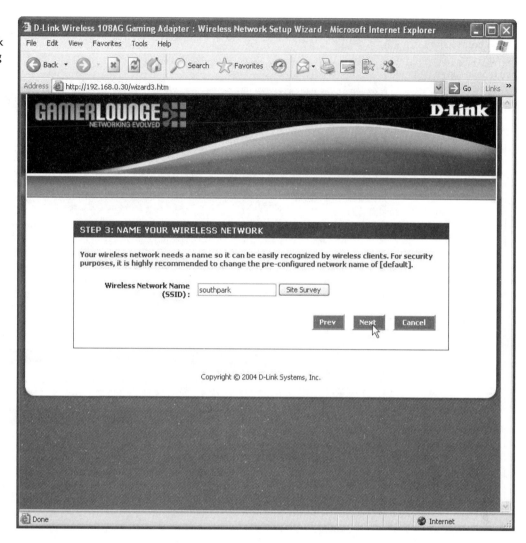

Step 2: Configure the Bridge to Be Part of Your Network

This step is a one-way street—after you complete it, the bridge and the computer connected to it become part of your existing network, and you will no longer be able to access the configuration web pages or change the settings of the bridge without resetting it.

In the LAN settings page of the bridge configuration, shown in Figure 20-9, select Enable DHCP and then click the Save Settings button. This setting will allow DHCP configuration requests to pass from your LAN's router/firewall to any computers connected to the other end of the bridge. Wait about 30–60 seconds for the gaming adapter to adjust to the new settings.

Figure 20-8

For the best security, use WPA and TKIP, and then type in the security key used by your wireless access point.

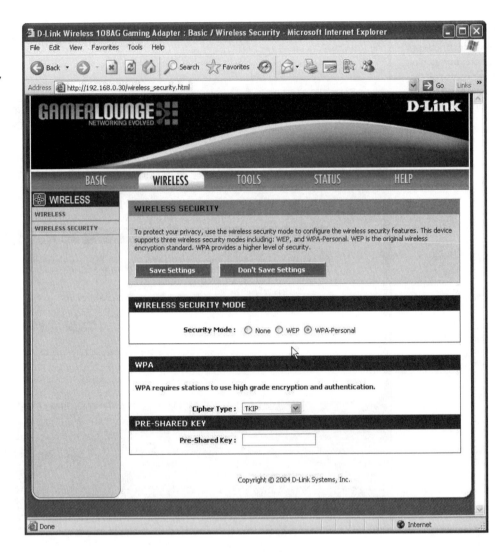

Step 3: Reset Your Computer to Work with the Bridge

We're going to reverse the computer IP address settings from Step 1 so that your computer connects to your existing network as before, through the bridge.

Change your computer's Ethernet address from 192.168.0.2 to DHCP. Click Start, right-click My Network Places, and choose Properties. Right-click your wired network adapter, choose Properties, and double-click the Internet Protocol (TCP/IP) entry. Select the Obtain An IP Address Automatically radio button, to allow your computer to pick up the IP address provided by the newly configured bridge. Click OK twice to close the dialog boxes and let the setting take effect.

In a matter of seconds, your computer and the network bridge will have adopted settings for your network, and you're ready to move the bridge to another part of your house to provide a wired connection to extend your network.

Figure 20-9

Enabling DHCP allows the bridge and the connected computer to adopt the settings of your existing network.

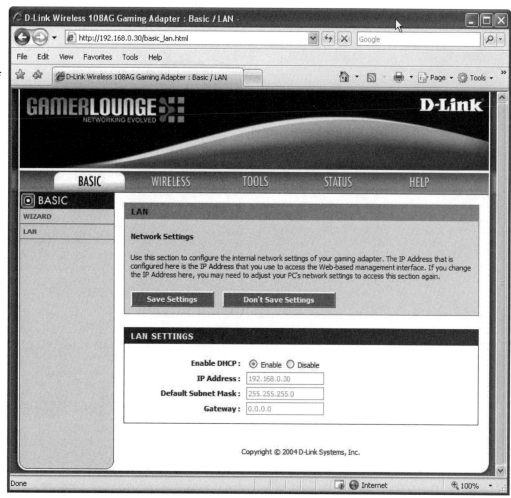

Expanding Beyond the Bridge

What You'll Need

- A wireless bridge—aka gaming adapter
- Leftover nonwireless router from Project 2
- Your computers with web browsers
- Cost: $0

Now that we've wirelessly jumped a huge span of real estate from one end of the house to the other in Project 20, we can reintroduce wires into our extended network configuration.

Our network continues to grow to include many computers at either end of a wirelessly bridged network, to the new network scheme of Figure 21-1. All we need is that old nonwireless router we removed from service in Project 2. Instead of being used as the main expansion of our one-connection DSL or cable service, and as a firewall, we reuse this gadget to become part of our local network at the far end of a wireless connection configured in Project 20. All this takes is a few simple reconfiguration steps to allow it to accept the IP address of our local network, turn off the firewall, and let it dish out a new set of IP addresses for the extended network clients.

Step 1: Reconfigure Your Wired Router

Our first step is to reconfigure our old wired router, the one we replaced with a wireless access point and router, to work at the end of the wireless connection we're making to expand our network:

1. Connect a computer to one of the LAN connections on the wired router using a single Ethernet cable. Leave the computer configured to obtain an IP address automatically (DHCP).

Figure 21-1

Schematic of the new
bridged network
to support multiple
computers at the
far end

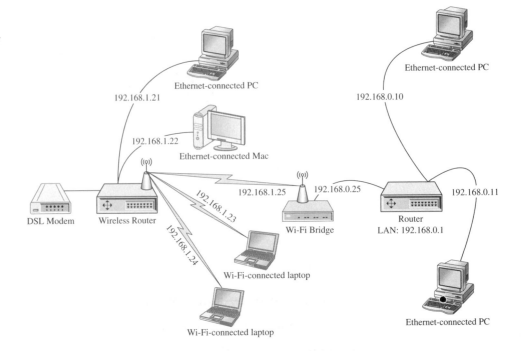

2. Access the router's web configuration page by entering the router's IP address into the Address bar of your web browser. Don't know what address to use? You could go to the Properties dialog box of the network adapter in your computer to find out, but an easier way is to use the Ipconfig program from a command prompt, as shown in Figure 21-2. Ipconfig quickly tells us the default gateway address is 192.168.0.1—this is the LAN address of the router. Put this address into your web browser.

Figure 21-2

The Ipconfig program
reveals IP address
information quickly.

```
C:\Documents and Settings\jim>ipconfig

Windows IP Configuration


Ethernet adapter Local Area Connection:

        Connection-specific DNS Suffix  . : Belkin
        IP Address. . . . . . . . . . . . : 192.168.0.36
        Subnet Mask . . . . . . . . . . . : 255.255.255.0
        Default Gateway . . . . . . . . . : 192.168.0.1

C:\Documents and Settings\jim>_
```

3. Access the Internet/WAN settings for the router, as shown in the example in Figure 21-3, and then select a Dynamic connection type and click Next. Your router will get a dynamic IP address through the bridge adapter we installed in Project 20. Click the Next button to access the final WAN configuration page, shown in Figure 21-4, skip entering a name, and click Apply Changes to convert your router from cable/DSL use to local network use.

Figure 21-3

Change your WAN type from PPPoE to Dynamic for your local network.

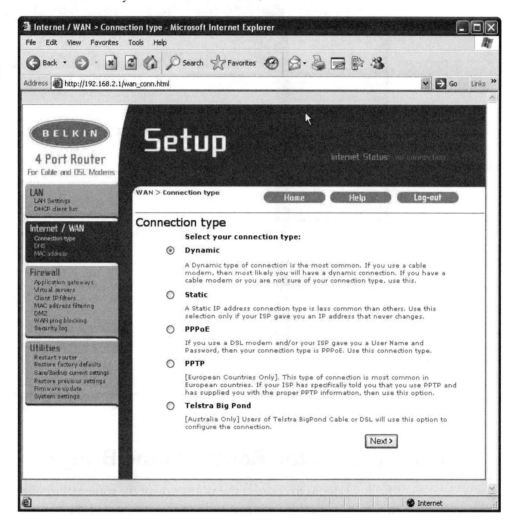

Step 2: Disable the Firewall

Since this project involves placing a router on an existing network that should have a firewall facing the Internet/WAN already, from Project 11, we do not need or want another firewall on our extended network. Access the Firewall web page for your router, as shown in the example in Figure 21-5, select Disable, and then apply the changes.

Figure 21-4

Apply the WAN type changes to your router.

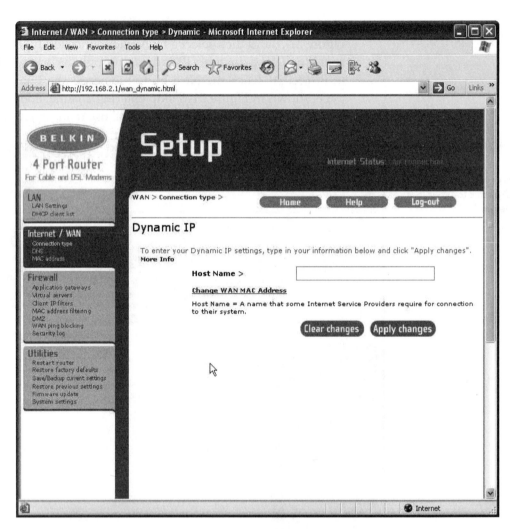

Step 3: Connect the Router to the Bridge

Our final step is to connect the WAN side of the router (the connection that used to go to the DSL or cable modem) to the Ethernet port of the gaming adapter. Disconnect and then reconnect the power at both units so that they reset and acquire new network addresses. After a minute or two you may connect computers to the LAN ports of the router and enjoy your expanded network connectivity.

In case you are wondering, yes, you could substitute another wireless router instead of using a wired-only router, or connect an access point to the LAN side of this router, extending your wirelessly bridged network even farther. If you choose to do this additional expansion, you must use a different Wi-Fi channel on this end from the one you used on the original wireless router of Project 2. You can use the same SSID name and security key for this access point so that your portable computers can roam from one end of your network to the other using the same settings.

Figure 21-5

Your extended network
does not require
another firewall, so this
one can be disabled.

Create a Server

What You'll Need

- **A spare computer**
- **Your local network**
- **Cost: Windows XP Home Edition or Professional operating system $129 (if needed)**

In previous projects we covered making Windows workgroups, sharing files and printers from computers used by other people, and controlling the desktop of one PC from another. These features are all very handy but have the drawback that we're using someone else's computer to do these things—and those users may get annoyed at the possible disruption of their work, games, and Internet use.

There is a way to avoid the negatives and enjoy the positives of sharing, by dedicating a separate PC to act as a server for all of these functions, and more. Then, if the "server" is Windows XP Professional or has a remote-control application installed, we can control it remotely instead of wasting space for an extra display, keyboard, and mouse. In the process we'll provide more granular control over the resources that different users can access, by turning off Simple File Sharing.

Step 1: Configure a PC to Be a Server

Start with a spare PC. Almost any PC will do, slow or fast, as long as it can run at least Windows XP Home Edition. It won't even have to run any applications, unless you want it to. We'll perform the familiar configuration steps to add the PC to our network as we did in Project 12, and some more advanced steps to make this PC behave more like a server than an ordinary PC or laptop.

1. Name this new PC something obvious as to its purpose. HOMESERVER works for me.

2. Place the PC in your named home workgroup (from Project 7)—perhaps it's JAKENET, named after your dog.

3. Configure a fixed IP address for this PC. To access these settings open Control Panel, open Network and Internet Connections. Click Network Connections. Right-click Local Area Connection then select Properties. Double-click Internet Protocol (TCP/IP).

4. If your router's DHCP range is from 192.168.1.2 through 192.168.1.31, use the last possible address in the range—it's unlikely another PC will snag the 30th available address unless you really have 30 PCs online at home at the same time. (Don't laugh, it's happened!)

5. Click the Advanced button then click the WINS tab. Make sure NetBIOS is enabled in your Advanced/WINS network settings. Click OK to close these dialogs.

6. Add and configure user accounts and their user levels—User, Power User, or Administrator as covered in Project 4.

7. Set up remote connection capabilities for this computer and configure which users you would like to be able to access it from their computer (from Project 14).

8. Install and share a common printer (from Project 4).

Step 2: Set Up Advanced File Sharing

In our file sharing project, Project 3, we could only share or not share folders, and any user who could access the shared folder could do everything—read, write, and delete—which perhaps is more control than you would like some users to have for most files. Advanced file sharing, or *not* Simple File Sharing, lets you not only control who can access a shared folder, but also grant very granular control over any aspect of file manipulation.

In Windows XP Professional, you simply choose Tools | Folder Options, click the View tab, deselect Use Simple File Sharing (Default), and click OK to close the dialog box. You then have access to a Security tab when viewing the Properties dialog box of a folder—shared or not.

By design, Windows XP Home Edition does not offer this same ability to turn off Simple File Sharing and allow you truly granular file and folder access control. That this feature is omitted by design does not mean that it is not possible to achieve the same granular control—you just have to work a little harder to achieve it. (If you have Windows XP Professional, you can skip to Step 3.)

The work-around is to enable the Security tab of the Customize Permissions dialog box, and then enable granular folder and file security. There are two methods to achieve this. The first method to enable the Security tab follows:

1. Go to ftp://ftp.microsoft.com/bussys/winnt/winnt-public/tools/scm/.

2. Download the SCESP4I.EXE file to your hard drive.

3. Locate and double-click SCESP4I.EXE to extract the files to a temporary location on your hard drive.

4. Go to the temporary location, right-click the setup.inf file, and choose Install.

5. When the installation is finished, reboot your computer.

6. Open Windows Explorer, open your C: drive, create a folder to share, right-click the folder, and choose Properties. You will see the Security tab of the Customize Permissions dialog box (shown later in Figure 22-3).

The following is the method of enabling granular file and folder security settings:

1. Open Windows Explorer, open your C: drive, and create a new folder to share.

2. Choose Start | Run.

3. Type **SHRPUBW** and then press the ENTER key or click the OK button.

4. Click the Browse button (see Figure 22-1) to choose the folder you want to share.

Figure 22-1

The SHRPUBW program is a poor-man's way around XP Home Edition's lack of granular security.

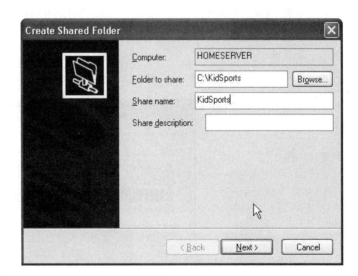

5. Type in a share name, and a share description if you like, and then click the Next button.

6. Choose the Customize Share and Folder Permissions radio button and click the Custom button (see Figure 22-2). The Custom dialog box of the Customize Permissions dialog box, shown in Figure 22-3 provides granular per-user control over files and folders.

Figure 22-2

Customize the shared
folder permissions to
manage granular folder
and file security.

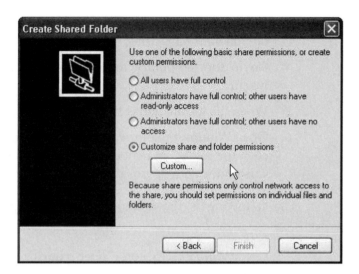

Use either of the preceding methods to create and configure your file shares, one
for music, one for photos, one for home business documents, one for homework, and
so on. Configure permissions and security for each shared folder (see Figure 22-3).
The following list explains the significance of each permission option:

- **Full control**   All rights in this list plus file deletion

- **Modify**   Open the file to change and write back onto the shared folder

Figure 22-3

Security rights to files
and folders provide
many options to protect
critical data files.

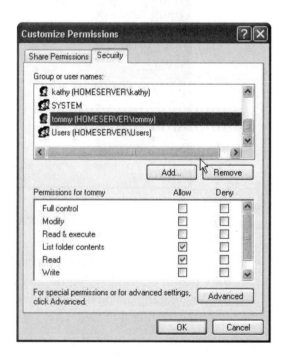

- **Read & execute**   View, copy, and, if an executable program or batch/command file, run the file

- **List folder contents**   Use Windows Explorer or the command line to see what files are in a folder

- **Write**   Create new files

- **Read**   View and copy the file

- **Special permissions**   Extended control over whether assigned rights can flow through to subfolders and files within them

For each of these security attributes, you have the option to Allow or Deny the level of file action, or to not set any specific permission and leave the security level set to the user's account level. If you simply do not set Allow rights (leaving both the Allow and Deny boxes unselected), the user attempting the interaction receives a message indicating they do not have the proper rights, if they can see the folder or files at all. If the level is set to Deny, the user receives a bit stricter message indicating they are clearly blocked from the interaction.

If you are really intent on denying any and all access to an available share, set both the permissions for the share to exclude or deny a specific user, and deny them every category of security access.

Step 3: Connect to the Server and Access Shared Resources

If you went through Projects 3, 4, 12, and 13, you've already done this part—simply follow the steps in those projects. There is one feature we did not cover in Project 3: making a shared drive on another computer or file server appear as just another drive letter to Windows Explorer and the command prompt. This is called *mapping* (or *associating*) an available drive letter to a storage resource. To map a shared network drive to a drive letter:

1. Double-click My Network Places on the Desktop or choose Start | My Network Places to open an Explorer window, and then, in the left navigation pane, click View Workgroup Computers to see the computers on your network.

2. Double-click the icon for the computer hosting the shared resource you want to use—in this case, your server, which we named HOMESERVER for this project. This will show you the available shared resources, as shown in Figure 22-4.

Figure 22-4

Viewing a workgroup
computer shows you
all available shared
resources.

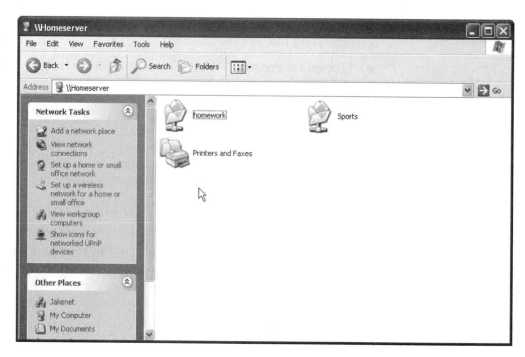

3. Locate the shared folder you want to turn into a drive letter for your local
   computer, right-click it, and choose Map Network Drive. In the Map Network Drive dialog box, shown in Figure 22-5, choose a drive letter to associate with the shared folder.

Figure 22-5

Selecting a drive letter
to associate with the
shared network folder

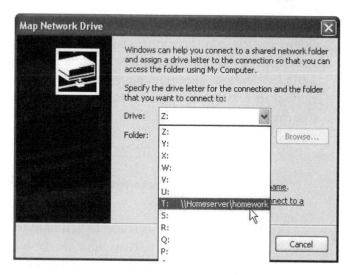

Repeat these drive-mapping steps for each shared folder you want to access as
a drive letter, on every computer on your home network that needs access to them.
With mapped drives and shared printers connected to your individual PCs, you've
just created a significant home server.

Project 23

Dealing with Dynamic IP Addresses

What You'll Need

- **A computer**
- **Your local network and Internet connection**
- **No-IP registration**
- **Cost: No-IP Plus service $24.95/year**

While we know and access most web sites by their fully qualified domain name, such as www.yahoo.com, www.espn.com, www.hgtv.com, and so forth, these are actually host+domain names that are translated by your ISP's Domain Name System server into numeric IP addresses. For a very large web site such as Yahoo.com, you are actually directed to one of several possible IP addresses, because a large web site divides the traffic among several different servers. You can witness this in practice by using the ping command to check the IP address of, say, www.yahoo.com one day, and then again another day; you'll find that the IP address changes, though you still get to the same web page from your browser.

A similar effect happens with your home cable or DSL Internet connection each time you connect and disconnect—but by different technology for different reasons. Therefore, it's not possible to expect that the IP address you give to your friends and family for your webcam to be useful tomorrow, next week, or next month.

To ensure that you have a consistent method to access services on your home network from across the Internet, you have two choices. The first is to change your cable or DSL service from a subscriber level with basic dynamic IP addressing to a premium and more expensive account with static IP addresses. The second is to sign up for a service that will automatically translate your dynamic IP address into a frequently updated fully qualified domain hostname—a domain name you register or a generic one.

Some of the most popular dynamic DNS services are No-IP from No-IP.com (www.no-ip.com), DynDNS (www.dyndns.com) from Dynamic Network Services, and DNS2Go (http://dns2go.com) from Deerfield.com. If you choose a free service, you get to pick a catchy hostname (nothing obvious like www) coupled with a choice of one of their top-level domain names. So, for example, you could end up with evil-pcgenius.no-ip.org to point to your home server. If you choose a paid service level, you could actually use a "nice" URL, like www.evilpcgenius.org.

The key to making this translation is a piece of software you run on one of your computers that checks for your current cable or DSL IP address and then sends that address to the dynamic IP provider's DNS servers and associates it with your host-name. The result is that your free URL evilpcgenius.no-ip.org or your registered do-main evilpcgenius.org could point to IP address 75.123.45.67 one day and the next day point to 71.98.76.54. As long as the DNS services for the ISPs of your friends and family catch the update from the dynamic DNS provider, your friends and family will be able to view your webcam or whichever service you run on your home network and allow in from the Internet.

Step 1: Register for a Dynamic IP Service

For some reason, long ago I chose one of No-IP.com's services for some of my dy-namic IP services, both free and a paid service, and just stuck with it, so that is what I'll use for this project. Visit www.no-ip.com (so-named for those of us who do not have static IP addresses) click the Sign-up Now link at the top of the page, create your account, and select your preference for hostname and their available domain names (see Figure 23-1).

1. At the No-IP web site click the Sign-up Now! link near the top of the web page.

2. Provide your information on the Create Your No-IP Account page. Read and accept the terms of use. You will be sent an account activation link to the e-mail address you provided.

3. Check your email account, click the link provided to activate your No-IP account, then log onto the site. Once you are logged in you can provide the details to add your desired hostname (Figure 23-1) for the No-IP service you select.

4. Next, select the Downloads tab at the top and click the respective Windows, Apple, or Linux logo to download the dynamic IP program for your com-puter. Install and configure the No-IP client program (Step 2 below) so it can update the No-IP DNS servers for your use.

Figure 23-1

Dynamic IP services let you be creative in your hostname selection and generic domain name.

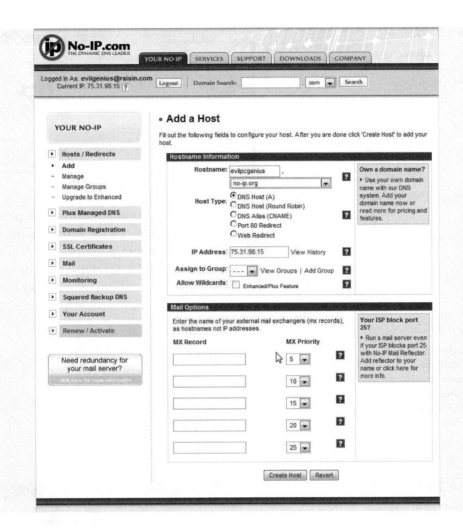

Step 2: Install the No-IP Dynamic Update Client Program

The No-IP Dynamic Update Client (DUC) program determines the IP address of your current broadband connection, logs onto your No-IP.com account, and sends the IP address update to associate the address with the host and domain names you selected for your home network applications. Each time you reset your broadband connection (or when your ISP dumps the connection) and it must be reconnected, the new address will be detected and updated to the No-IP.com servers.

1. Locate and run the Ducsetup.exe program to begin the No-IP DUC installation. The first screen, shown in Figure 23-2, advises you that your computer must be connected to the Internet. Click the Next button and accept or change the default settings in the next dialog box, and then click Next again.

Figure 23-2

An Internet connection is required for the No-IP DUC program.

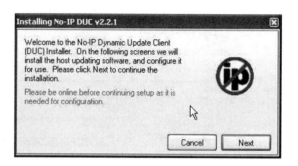

2. Provide your No-IP registration e-mail address and password (see Figure 23-3) so that No-IP DUC can connect, verify your credentials, and associate your current IP address with your choice of host and domain names. When your credentials are verified, No-IP DUC sends your current IP address to the No-IP.com servers and shows you the hostnames you have just associated with your IP address, as shown in the example in Figure 23-4.

Figure 23-3

Your dynamic IP address and hostname association are protected by your username and password.

Figure 23-4

The No-IP DUC screen is the place to check for your IP address and hostnames.

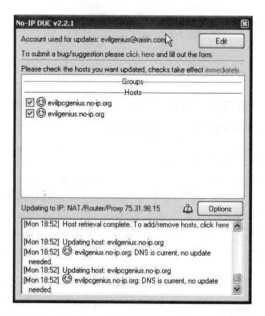

3. Click the Options button to configure No-IP DUC to automatically run at startup or as a system service (in the background), as shown in Figure 23-5, and to verify other settings.

Figure 23-5

The No-IP DUC
options let you select
how and when to run
the program and update
your IP address.

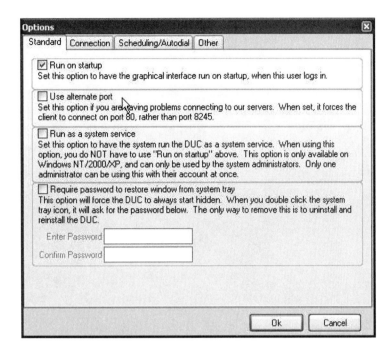

With these steps completed, you can now tell your friends to enter your new host-name into their browsers, along with the port number for your webcam server (from Project 6), and enjoy the view without those messy IP addresses.

This project works for hosting your own web pages (if allowed by your ISP), file transfers, remote-control sessions, and many other Internet-to-home connections.

Access TiVo
Files Anywhere

What You'll Need

- **Setup from Project 8**
- **Dynamic DNS service to be able to identify and find your home Internet connection**
- **Cost: $0**

You probably know you can schedule TiVo recordings through TiVo Central Online, but you may not know that you can download your recorded programs to a TiVo Desktop anywhere on the Internet. Of course, TiVo recordings are huge files, and most home DSL connections can upload at only 384 to 512 Kbps, so you may want to reduce the recording quality you use on your TiVo DVR so that the transfers take less time.

This project involves tweaking your home router/firewall to pass requests for access to your TiVo from the Internet to the internal web pages of your TiVo.

Step 1: Verify Access to Your TiVo DVR Built-In Web Page

In this step you will verify you can access the web server and file selections built into your TiVo DVR.

1. Begin with the IP address assigned to your TiVo DVR in Project 8. Type the IP address, preceded by https:// into your web browser's address bar

    ```
    https://your TiVo's IP address
    ```

    If your TiVo's internal web page is active—and it probably is if you use the TiVo Desktop—you first encounter a server certificate error or warning message, shown in Figure 24-1. You need to accept the certificate, either permanently or temporarily, to gain further access.

Figure 24-1

The TiVo internal web pages are not a typical Internet site but the security certificate can be trusted.

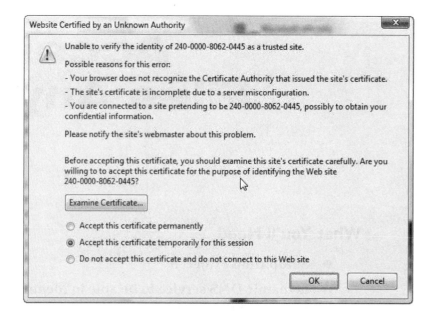

2. In the browser pop-up that appears type in **tivo** as the user name and the password, which is the Media Access Key of your TiVo unit.

3. Click OK and you'll find yourself at a web page listing your Now Playing list, as shown in Figure 24-2.

Figure 24-2

TiVo's built-in Now Playing list web page

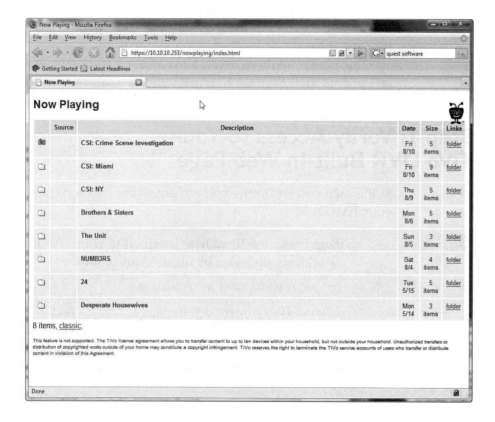

4. Select a folder in the Links column to the right and you'll see all of the recordings for one of your shows, as shown in the example in Figure 24-3.

Figure 24-3

Expanding a selection from TiVo's Now Playing page lists the shows saved on the TiVo hard drive.

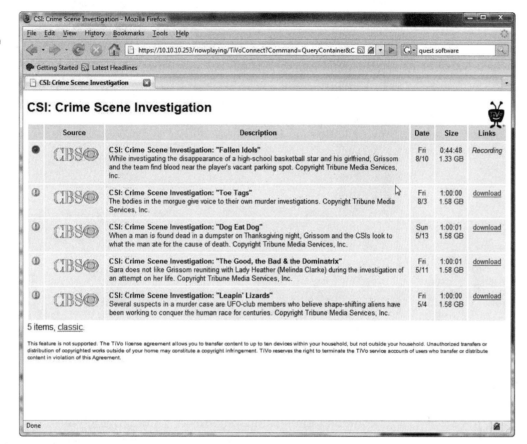

5. Select a download link to the right and you'll be able to save a .Tivo file to your computer that can be played on a computer that has the TiVo Desktop installed, which includes the file converter for TiVo files to play them through Windows Media Player or Apple QuickTime. Yes, you can do this with the TiVo Desktop software, but this is a test for the next step—to make it possible to do this over the Internet.

Step 2: Adjust Your Router Settings to Connect to TiVo over the Internet

This step requires you to revisit Project 3 to configure your broadband firewall to allow inbound connections from the Internet to your TiVo.

Access your firewall, and navigate to the pinhole, port forwarding, or special application settings page to add the TiVo address and TCP port 443 to the inbound access list (see Figure 24-4 for an example).

Figure 24-4

Configure your router's firewall to allow access to TiVo's secure web page on LAN port 443.

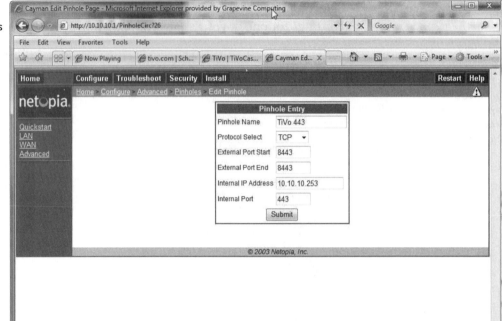

In this example, you will see that I had to configure the external port setting to 8443 to get past ISP restrictions against using port 443, and then have it forwarded to port 443 to the TiVo's IP address.

After this, you will want to use a dynamic IP service such as No-IP.com and its local computer application (see Project 23) to provide you with a steady hostname to use over the Internet to access your dynamically addressed cable or DSL connection.

You may be asking why you cannot simply connect to your TiVo over the Internet using the TiVo Desktop to download or share files. An excellent question, indeed. The simple answer is that the TiVo Desktop software does not allow you to provide an IP address or hostname to find your TiVo with; instead, it sends a beacon from a computer for the TiVo to find—which does not work well because thousands of other TiVos will also find the beacon. The more complicated answer is that you can connect to your TiVo over the Internet using the TiVo Desktop to download or share files, but doing so requires a variety of not-so-simple software and some hacking inside the TiVo computer, as described in the book *TiVo Hacks* by Raffi Krikorian (O'Reilly Media, 2003) and in dozens of techie blogs.

Step 3: Reconfigure Firewall Settings for TiVo Desktop

You need to reconfigure your Internet firewall settings to allow the following ports access between your computers and TiVo so that the TiVo Desktop server or web browser can communicate with your TiVo on your local network. The web browser needs only TCP port 443, while the full TiVo Desktop software requires several network port connections, as follows:

- TCP port 37
- TCP port 2190
- TCP port 4430
- TCP port 7287
- TCP port 7288
- TCP port 8000
- TCP ports 8080–8089
- UDP port 123
- UDP port 443
- UDP port 2190

Index

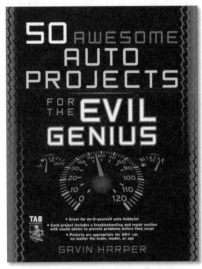

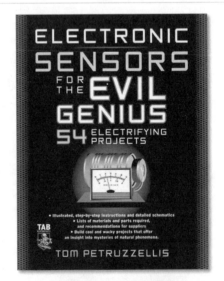

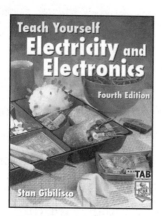

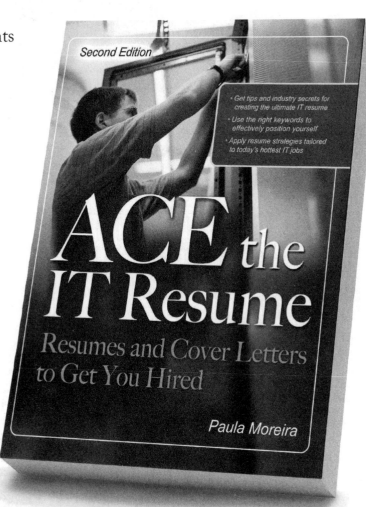